# LOGIC AND ALGORITHMS

# Logic and Algorithms

## With Applications to the
## Computer and Information Sciences

**ROBERT R. KORFHAGE**

Assistant Professor of Mathematics and Computer Sciences
Purdue University

**John Wiley & Sons, Inc., New York · London · Sydney**

A /164

Library of Congress Catalog Card Number: 66-25225
Printed in the United States of America

To Bernard A. Galler
Friend and guide
in this MAD world

# Preface

This book has evolved from the basic logic course for students in mathematics and the computer sciences at Purdue University, and a similar but more elementary course presented to an industrial group of computer users who in general did not have much of a mathematical background beyond that acquired in high school.

In developing these courses I discovered that logic books might be grouped into three main categories: those written by philosophers, who are inclined to emphasize the meaning and use of logic in ordinary discourse; the mathematical logic books, which brush aside such interpretations and concentrate on a formal, algebraic approach; and the books written for engineers, which are concerned with the applications of Boolean algebra to switching theory and make but slight reference to logic in any other context.

In the field of algorithm theory, the situation has been especially dim: only one elementary book is available.

This book is written on the premise that computer users and designers need to have some knowledge of all three aspects of logic, as well as the fundamentals of algorithm theory. We are concerned ultimately with a communication problem—among people, or between people and computers. The topics covered have been chosen with this in mind. First, the elements of logic which we shall discuss are themselves important for precision in communication. But of more significance, they provide a simple yet sophisticated example of a formal symbol manipulation or communication system. It should be remembered throughout the book that although we are discussing logic, the same general types of symbol manipulation rules and techniques are used for handling all symbols, be they logical, mathematical, chemical, linguistic, or whatever.

The idea of a set forms a unifying concept for modern mathematics, and is the underlying basis of this book. Thus the first chapter covers the basic concepts involved in set algebras, relations, and mappings.

Boolean algebras constitute a simple abstraction of set algebras, and are discussed next. Not only are these algebras of importance in the design of digital computers, but also they provide a bridge between set theory and elementary logic.

The subject of logic itself is divided between Chapters 3 and 6, with the propositional calculus in the first of these chapters and quantification theory in the latter. Most of the material in these chapters is nonaxiomatic, although appropriate axiom systems are introduced and discussed toward the end of each chapter.

We often hear novices in the computing field express some puzzlement that the computer—supposedly a numerical device—can "read" alphabetic characters. In Chapter 4 I have tried to point out the abstract nature of the internal symbol representations of a computer by examining several possible interpretations of a binary vector.

The chapter on algorithms or methods of solving problems centers initially on the work of Turing and Markov, for it is felt that these two approaches bring forth most clearly the concepts involved. However, since Turing machines and Markov algorithms are *in practice* poor computational tools, I have also provided examples of algorithms written in the languages used in computing—flowcharts, a rudimentary assembly language, and the MAD language.

In line with the philosophy that our main problem is one of communication, these various topics are tied to each other and to linguistics through the medium of formal languages. Finally, a brief history of the development of the subjects covered in this book is included.

The amount of material which might be covered in a one-semester course using this book as a text will vary with the mathematical sophistication of the students. It is not that a full knowledge of other branches of mathematics is necessary, but rather that students with more of a mathematical background are more capable of grasping the flow of the arguments, and hence they can absorb the material faster. I have found that the material in the first six chapters is more than enough for a nonmathematical group to handle. On the other hand, in teaching a group of seniors and graduate students in mathematics and computer science I am able to cover all the material given here and still spend several weeks on axiomatic logic. Thus I feel that the book by itself provides an adequate text for a course aimed at sophomores and juniors.

The inclusion of answers to almost all the exercises should prove helpful to the person who wishes to study on his own. It should be noted, however, that the answers given for problems involving proofs or the design of algorithms or Turing machines are not the only possible solutions to the exercises.

I would like to acknowledge the helpful comments in the development of this book which have been made by M. Drazin, R. E. Miller, D. Muller, R. W. Ritchie, and other colleagues and students.

Robert R. Korfhage

*May 1966*

# Contents

# Special Symbols

# 1. Sets, Relations, and Mappings

## 1. INTRODUCTION AND NOTATION

In order to facilitate our overall discussion, we must first define some basic terms and concepts. In this chapter we do so with the introduction of sets and the associated operations on sets. In addition we develop an algebra of sets which will provide concrete examples of the Boolean algebras to be discussed in Chapter 2.

One point on which many formal and informal discussions of any topic may collapse is the matter of defining basic terms. "Co-existence" may mean one thing to us and another to the Russians; but unless there is a common understanding of its meaning, any discussions of its implications are meaningless. However, in trying to define this or any other term, words must be used—words which themselves must be defined in terms of other words, and so on. Clearly, this process must ultimately stop at some words which could be, but are not (by common consent) defined: there must be some undefined, but mutually understood, terms.

So also in mathematics. We shall take as our undefined terms *set* and *element*, and the phrase *is an element of*. Thus we may discuss these words and say informally what we intend their meaning to be, but we shall not define them.

We intend that a set $A$ is a collection of objects, called the elements of the set $A$. (Note that it does not help us to *define* a set as a "collection of objects," for then we must say what we intend the phrase "collection of objects" to mean.) These objects may be anything: concepts, symbols, letters, physical objects, sets, etc. We shall assume that whenever we write "$x$ is an element of $A$" we could also write "$x$ is an element" and "$A$ is a set." The only restriction which we place is that it be meaningful to ask "Is $x$ an element of $A$?", and that for each $x$ and each $A$ the question may in some way be answered "yes" or "no."

*Example 1.1.* The following are typical sets.

The set of all positive integers.
The set of all fire hydrants in New York City.
The set of all English words beginning with "$a$".
The set of all colorless green ideas.
The set of all sets of integers.

1

It is well known that a formal set theory built on this broad framework leads to paradoxical situations. Thus if one desires to construct a set theory one must avoid allowing such things as "the set of all sets" to be a set. Since our primary interest is not in this direction, we will not go into the complexities of avoiding the paradoxes, but will assume that the sets we have in mind are well-behaved.*

We shall adopt the following notational scheme for the discussion of sets. Sets will be denoted by capital letters $A, B, C, \ldots$; elements will be denoted by lower case letters $a, b, c, \ldots$; and the phrase "is an element of" will be denoted by the symbol "$\in$". Thus we write "$x \in A$" for "$x$ is an element of $A$." Here, and in analogous situations, we indicate the denial of a statement by use of "/", writing "$x \notin A$" for "$x$ is not an element of $A$." In addition, we shall use the word "member" as synonymous with "element."

There are two basic ways to describe a set: either by listing the elements of the set, or by describing a property which holds for all members of the set and only for those objects. For a "small" finite set, an enumeration or listing of the elements is easily accomplished. For example, if $1 \in A$, $\# \in A$, and $@ \in A$, and these are the only elements of $A$, we may write $A = \{1, \#, @\}$, where the braces are the conventional set delimiters. For a "large" finite set, say one with a million elements, such an enumeration is still possible, yet clearly more tedious. And of course for a set with infinitely many elements, a complete enumeration is impossible. Here one finds commonly the use of ellipses ($\ldots$) to indicate the unlisted elements. Although this notation is imprecise, within a given context the meaning is generally clear.

*Example 1.2.* The use of ellipses is illustrated by the following sets.

The set of all letters of the English alphabet—$\{a, b, \ldots, z\}$.

The set of all positive even integers—$\{2, 4, 6, \ldots\}$.

It should be noted that the order of enumeration is not significant— $\{1, \#, @\}$ is the same set as $\{@, \#, 1\}$ or $\{\#, @, 1\}$. Also the multiple listing of a single element in a set is without meaning. Thus the set $\{1, 1, 2, 3\}$ is the same as the set $\{1, 2, 3\}$. If it is really intended that the set contain two or more objects which are indistinguishable then these objects should be labeled, as $\{1, 1', 2, 3\}$.

Whenever an enumeration is difficult or impossible to obtain, either because of the number of elements in the set or because of other reasons (for example, the set of all three-legged dogs in the United States), it is often sufficient to describe the set by a characteristic property. The notation generally used for this is $A = \{x \mid P(x)\}$; that is, $A$ is the set of all elements $x$ having property $P$ (or such that property $P$ holds).

* For further information on the paradoxes, see Rosenbloom [2] or Wilder [5].

*Example 1.3.* These sets illustrate this mode of set description:

The set of all even integers—{$x \mid x$ is an integer and $x/2$ is an integer}.

The set of all words in this book—{$x \mid x$ is a word and there is at least one occurrence of $x$ in the book *Logic and Algorithms*, by R. R. Korfhage}.

### EXERCISES

1. Describe each of the following sets in words.
    (a) {$a, b, c, d, e$}
    (b) {1, 5, 10, 25, 50, 100}
    (c) {$x \mid$ for some integer $y$, $x = 3y$}
    (d) {$x \mid x$ is a planet in the solar system, having more than two moons}.
2. Write each of the following sets in standard notation.
    (a) The set of all second letters of English words (not including proper names) beginning with "y".
    (b) The set of all months in the present year containing a Friday the 13th.
    (c) The set of all numbers whose squares are positive (that is, greater than zero).
    (d) The set whose elements are all sets whose elements are elements of {( , )}.

## 2. SOME SPECIAL SETS

Because the elements which may (or may not) be in a set may be any conceivable type of objects, we find it desirable to introduce a *universal set*. If we consider for example $A = \{x \mid 2x = 5 \text{ or } x = 0\}$, it is clear that $A$ is intended to be a set of numbers. On the other hand the symbolism $\{x \mid 2x = 5 \text{ or } x = 0\}$ does not say that $x$ is a number: one might conceivably search for an $x$ satisfying $2x = 5$ among such things as triangles, green fire hydrants, and abstract paintings. It is of course permissible to write $A = \{x \mid x \text{ is a number, and } 2x = 5 \text{ or } x = 0\}$, but if one is dealing with many sets of numbers this is cumbersome and repetitious. Thus one makes a statement such as "the universal set $U$ is the set of numbers" and then tacitly assumes that any element referred to has the property that it is an element of the particular prescribed universal set. Thus a universal set does not contain everything, but rather is just the universe of discourse at a particular time.

There is also another reason for insisting on the specification of a universal set, namely that the elements of a set depend on the universal set. Thus for $A = \{x \mid 2x = 5 \text{ or } x = 0\}$, if $U$ is the set of all real numbers,

$A = \{0, 2\frac{1}{2}\}$; but if $U$ is the set of all integers, $A = \{0\}$, since $2\frac{1}{2}$ is not an integer.

To continue with this example, suppose that we take as the universal set the set of all odd integers. Clearly, 0 is not an odd integer nor is there any odd integer satisfying $2x = 5$. Thus $A$ has no elements. Hence we have need of another special set $\varnothing$, called the *null set* or *empty set*, which is the set containing no elements. This may be characterized as $\{x \mid x \neq x\}$.

Often one is interested in only certain elements of a given set. For example, one man may be interested in a book on protein chemistry, another in a book on physical chemistry. Taking the set of all books in a given library as a universal set, both men are interested in the set of chemistry books. Yet each man is interested in a different portion or subset of this set.

**Definition 2.1.** We say that $A$ is a *subset* of $B$, written $A \subseteq B$, if and only if every element of $A$ is also an element of $B$; that is, whenever $x \in A$, then also $x \in B$.

Generally the property of being a subset involves an extension of the definitional properties: if $A \subseteq B$, then every element of $A$ has those properties which place it in $B$, and in addition certain other properties which distinguish it from those elements of $B$ which are not elements of $A$. For example, a book on protein chemistry is a chemistry book, but it is also on *protein* chemistry as opposed, say, to physical chemistry.

We note in particular that a set $A$ is always a subset of itself, since whenever $x \in A$, then $x \in A$. The empty set is also a subset of any given set $A$, since the condition of the definition is vacuously satisfied. To use a more detailed argument, if $\varnothing$ is not a subset of $A$, then there must be some $x \in \varnothing$ such that $x \notin A$. But this is impossible since $\varnothing$ has no elements. Thus $\varnothing \subseteq A$.

**Definition 2.2.** Two sets $A$ and $B$ are *equal*, $A = B$, if and only if $A \subseteq B$ and $B \subseteq A$. (That is, every element of $A$ is an element of $B$, and vice versa.)

*Example 2.1.* Taking $U$ as the set of real numbers, if $A = \{1, 2\}$ and $B = \{x \mid x^2 - 3x + 2 = 0\}$, then $A = B$.

**Definition 2.3.** $A$ is a *proper subset* of $B$, $A \subset B$, if $A \subseteq B$ and $A \neq B$.

We are frequently interested in objects not having a certain property. For example we might be interested in books on protein chemistry, but not those written before 1960. Thus there is a subset $E$ of the set of chemistry books, namely those written before 1960, in which we are not interested. Of course the books which are not in $E$ themselves form a set, namely the set of chemistry books which are not in $E$.

**Definition 2.4.** Let $U$ be the universal set and $A$ any set. The *complement* of $A$, $\overline{A}$, is defined as $\{x \mid x \notin A\}$ (or more fully $\{x \mid x \in U$ and $x \notin A\}$).

If $A$ and $B$ are sets, the *relative complement of $A$ with respect to $B$* is
$B - A = \{x \mid x \in B \text{ and } x \notin A\}$.

  *Example 2.2.* If $U = \{0, 1, 2, 3, 4, 5, 6, 7, 8\}$, $A = \{1, 4, 6, 7\}$,
$B = \{1, 2, 4, 5, 7\}$, and $C = \{2, 5\}$, then $\bar{A} = \{0, 2, 3, 5, 8\}$, $\bar{B} = \{0, 3, 6, 8\}$,
$\bar{C} = \{0, 1, 3, 4, 6, 7, 8\}$, and the relative complement $\alpha - \beta$ is given by
Table 2.1.

**Table 2.1**

$\alpha - \beta$, Example 2.2

| $\alpha \backslash \beta$ | $A$ | $B$ | $C$ |
|---|---|---|---|
| $A$ | $\varnothing$ | $\{6\}$ | $A$ |
| $B$ | $C$ | $\varnothing$ | $\{1, 4, 7\}$ |
| $C$ | $C$ | $\varnothing$ | $\varnothing$ |

EXERCISES

1. Determine the elements of the set $A = \{x \mid x^2 = 11x - 30 \text{ or } 4 - x > 0\}$
   when the universal set $U$ is
   (a) the set of real numbers,
   (b) the set of rational numbers,
   (c) the set of integers,
   (d) the set of positive integers,
   (e) the set of negative integers,
   (f) the set of odd integers,
   (g) the set of even integers,
   (h) the set of integers greater than 10.
2. Let $U = \{0, 1, 2, \ldots, 9, 10\}$. Determine the complement of each of the
   following sets.
   (a) $\{1, 2, 4, 6, 7, 10\}$
   (b) $\{1, 4, 7, 8, 9\}$
   (c) $\{0\}$
   (d) $U$
   (e) $\varnothing$
3. Let $U = \{a, e, i, o, u\}$. Determine the complement of each of the
   following sets.
   (a) $\{a, e, u\}$
   (b) $\{i, o, u\}$
   (c) $U$
   (d) $\varnothing$

4. For each of the following, determine the relative complement $A - B$.
   (a) $A = \{1, 4, 7, 10\}$          $B = \{1, 2, 5\}$
   (b) $A = \{1, 2, 5\}$          $B = \{1, 4, 7, 10\}$
   (c) $A = \{a, e, i\}$          $B = \{e, a, i\}$
   (d) $A = \{a, ), 17\}$          $B = \{)\}$
   (e) $A = \{1, 5, 6, a\}$          $B = \varnothing$
   (f) $A = \{2, 4\}$          $B = \{1, 2, 4, 7\}$
   (g) $A = \varnothing$          $B = \{a, ), 17\}$

5. In each of the following, determine which of the two sets $A$ and $B$ is a subset of the other.
   (a) $A = \{1, 4, 7\}$          $B = \{4, 10, 1, 7\}$
   (b) $A = \{2, 5, a, 6\}$          $B = \{a\}$
   (c) $A = \{1, 2, a, b\}$          $B = \{a, 1, b, 2\}$
   (d) $A = \varnothing$          $B = \{a, b, c\}$
   (e) $A = \varnothing$          $B = \varnothing$
   (f) $A = \{1, a, 7\}$          $B = \{a, b, 7\}$

### 3. COMBINATIONS OF SETS

In this section we wish to discuss the formation of new sets from given ones, building towards an algebra of sets. We introduced one way of forming new sets—through forming the complement or relative complement—in the last section. We will now define three new operations and examine the relationships which exist between these various operations.

**Definition 3.1.** Let $A$ and $B$ be any sets. The *union* of $A$ and $B$ is $A \cup B = \{x \mid x \in A$ or $x \in B$ or both$\}$. The *symmetric difference of $A$ and $B$* is $A \triangle B = \{x \mid x \in A$ or $x \in B$ but not both$\}$. The *intersection* of $A$ and $B$ is $A \cap B = \{x \mid x \in A$ and $x \in B\}$.

*Example 3.1.* Let us define $U$, $A$, $B$, and $C$ as in Example 2.2. The union, symmetric difference, and intersection are given in Tables 3.1, 3.2, and 3.3, respectively.

**Table 3.1**

$\alpha \cup \beta$, Example 3.1

| $\alpha \backslash \beta$ | $A$ | $B$ | $C$ |
|---|---|---|---|
| $A$ | $A$ | $\{1, 2, 4, 5, 6, 7\}$ | $\{1, 2, 4, 5, 6, 7\}$ |
| $B$ | $\{1, 2, 4, 5, 6, 7\}$ | $B$ | $B$ |
| $C$ | $\{1, 2, 4, 5, 6, 7\}$ | $B$ | $C$ |

**Table 3.2**

$\alpha \triangle \beta$, Example 3.1

| $\alpha\backslash\beta$ | $A$ | $B$ | $C$ |
|---|---|---|---|
| $A$ | $\varnothing$ | {2, 5, 6} | {1, 2, 4, 5, 6, 7} |
| $B$ | {2, 5, 6} | $\varnothing$ | {1, 4, 7} |
| $C$ | {1, 2, 4, 5, 6, 7} | {1, 4, 7} | $\varnothing$ |

**Table 3.3**

$\alpha \cap \beta$, Example 3.1

| $\alpha\backslash\beta$ | $A$ | $B$ | $C$ |
|---|---|---|---|
| $A$ | $A$ | {1, 4, 7} | $\varnothing$ |
| $B$ | {1, 4, 7} | $B$ | $C$ |
| $C$ | $\varnothing$ | $C$ | $C$ |

It is helpful to draw diagrams, called *Venn diagrams*, to visualize these relationships. In Figure 3.1 the rectangle represents the universal set and the two circles represent sets $A$ and $B$. The various regions are numbered so that we may easily indicate the various relationships.

Venn diagrams may be used to demonstrate the basic relationships which hold between the set operations, as is shown in this demonstration that $A \cap (B \cup C) = (A \cap B) \cup (A \cap C)$. In Figure 3.2 we see that $A$ consists of the regions 2, 3, 4, and 5, and that $B \cup C$ consists of regions 3, 4, 5, 6, 7, and 8. Thus $A \cap (B \cup C)$ (the part in both $A$ and $B \cup C$) consists of regions 3, 4, and 5. On the other hand, $A \cap B$ consists of regions 3 and

**Table 3.4**

The sets in Figure 3.1

| Set | Region | Set | Region |
|---|---|---|---|
| $U$ | 1, 2, 3, 4 | $A \cup B$ | 2, 3, 4 |
| $A$ | 2, 3 | $A \triangle B$ | 2, 4 |
| $B$ | 3, 4 | $A \cap B$ | 3 |
| $\overline{A}$ | 1, 4 | $A - B$ | 2 |
| $\overline{B}$ | 1, 2 | $B - A$ | 4 |

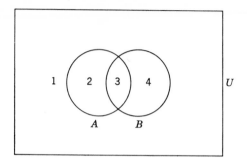

Figure 3.1. Venn diagram for two sets.

4, and regions 4 and 5 form $A \cap C$. Thus $(A \cap B) \cup (A \cap C)$ (those regions in either $A \cap B$ or $A \cap C$) consists of regions 3, 4, and 5. Hence the two expressions designate the same set.

While Venn diagrams are useful aids to understanding the basic set operations, their use is limited. The diagrams for four or five sets are quite complicated, and beyond that they become quite hopelessly tangled. Fortunately there is no need to consider more than three sets at a time in order to understand the fundamentals.

EXERCISES

1. Find the union of each of the following pairs of sets.
   (a) $A = \{2, 5, 8\}$                $B = \{1, 2, 5, 10\}$
   (b) $A = \{1, 2, 4\}$                $B = \{1, 2, 4, 7\}$
   (c) $A = \{a, *, \$\}$               $B = \{1, 2, 3\}$
   (d) $A = \varnothing$                $B = \{a, *, \$\}$
   (e) $A = \{1, 3, 5\}$                $B = \{2, 4, 6\}$

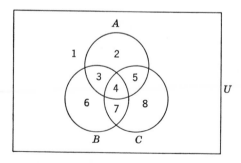

Figure 3.2. Venn diagram for three sets.

2. Find the symmetric difference of each of the pairs of sets in Exercise 1.
3. Find the intersection of each of the pairs of sets in Exercise 1.
4. Use Venn diagrams to verify each of the following relationships.
   (a) $A \cap (B \cap C) = (A \cap B) \cap C$
   (b) $A \cup (B \cap C) = (A \cup B) \cap (A \cup C)$
   (c) $A \cap (B \cup C) = (A \cap B) \cup (A \cap C)$
   (d) $A \cap (B \triangle C) = (A \cap B) \triangle (A \cap C)$
   (e) it is not true that $A \triangle (B \cap C) = (A \triangle B) \cap (A \triangle C)$
   (f) $\overline{A \cap B} = \overline{A} \cup \overline{B}$
5. $A$ is a subset of $B$ if and only if $A \cup B = B$, as may be verified by Venn diagrams. Determine expressions using the intersection, relative complement, and symmetric difference which hold if and only if $A$ is a subset of $B$.

## 4. SET ALGEBRA

As we work with the set operations, various properties of these operations appear. For example, from the above examples it appears that union, intersection, and symmetric difference are *commutative* operations: that is, $A \cup B = B \cup A$, $A \cap B = B \cap A$, and $A \triangle B = B \triangle A$; and one may demonstrate with Venn diagrams that this is indeed the case. On the other hand the relative complement is generally not commutative: $A - B \neq B - A$. Some other properties which may be demonstrated are given in Table 4.1. In the last line of the table, $\overline{\overline{A}}$ denotes the complement of $\overline{A}$.

### Table 4.1

Basic set operations

| | | |
|---|---|---|
| $A \cup U = U$ | $A \triangle U = \overline{A}$ | $A \cap U = A$ |
| $A \cup A = A$ | $A \triangle A = \varnothing$ | $A \cap A = A$ |
| $A \cup \overline{A} = U$ | $A \triangle \overline{A} = U$ | $A \cap \overline{A} = \varnothing$ |
| $A \cup \varnothing = A$ | $A \triangle \varnothing = A$ | $A \cap \varnothing = \varnothing$ |

$$A \cup B = (A \triangle B) \cup (A \cap B) = (A \triangle B) \triangle (A \cap B)$$
$$A \triangle B = (A \cup B) - (A \cap B)$$
$$\overline{A \cup B} = \overline{A} \cap \overline{B}$$
$$\overline{A \cap B} = \overline{A} \cup \overline{B}$$
$$A \cap (B \cup C) = (A \cap B) \cup (A \cap C)$$
$$A \cup (B \cap C) = (A \cup B) \cap (A \cup C)$$
$$\overline{\overline{A}} = A$$

It is useful to systematize our knowledge of these properties in the form of an *algebra of sets*. There is a parallel here to the usual algebra of numbers, although we shall see that the analogy is imperfect. When dealing with numbers (either rational or real) the two "basic" operations are addition and multiplication, and they satisfy certain properties:

Let $a$, $b$, and $c$ denote any numbers.

*Closure:*  **A1.**  $a + b$ is a number.

**M1.**  $a \cdot b$ is a number.

*Associativity:*  **A2.**  $(a + b) + c = a + (b + c)$.

**M2.**  $(a \cdot b) \cdot c = a \cdot (b \cdot c)$.

*Identity:*  **A3.**  There is a unique number 0 such that for any number $a$, $a + 0 = 0 + a = a$.

**M3.**  There is a unique number 1 such that for any number $a$, $a \cdot 1 = 1 \cdot a = a$.

*Inverse:*  **A4.**  For each number $a$, there is a unique number $-a$ such that $a + (-a) = (-a) + a = 0$.

**M4.**  For each number $a \neq 0$, there is a unique number $a^{-1}$ such that $a \cdot a^{-1} = a^{-1} \cdot a = 1$.

*Commutativity:*  **A5.**  $a + b = b + a$.

**M5.**  $a \cdot b = b \cdot a$.

*Distributivity:*  **D1.**  $a \cdot (b + c) = a \cdot b + a \cdot c$.

**D2.**  $(a + b) \cdot c = a \cdot c + b \cdot c$.

The analogy between ordinary algebra and set algebra is most nearly complete if we use the operation of symmetric difference ($\triangle$) in place of addition, and intersection ($\cap$) in place of multiplication. In fact, all of the above properties then hold except M4. The "unique number 0" becomes the "unique set $\varnothing$," and the "unique number 1" becomes the "unique set U."

**Table 4.2**

$\alpha \triangle \beta$

| $\alpha \backslash \beta$ | $\varnothing$ | $A$ | $B$ | $C$ | $D$ | $E$ | $F$ | $U$ |
|---|---|---|---|---|---|---|---|---|
| $\varnothing$ | $\varnothing$ | $A$ | $B$ | $C$ | $D$ | $E$ | $F$ | $U$ |
| $A$ | $A$ | $\varnothing$ | $D$ | $E$ | $B$ | $C$ | $U$ | $F$ |
| $B$ | $B$ | $D$ | $\varnothing$ | $F$ | $A$ | $U$ | $C$ | $E$ |
| $C$ | $C$ | $E$ | $F$ | $\varnothing$ | $U$ | $A$ | $B$ | $D$ |
| $D$ | $D$ | $B$ | $A$ | $U$ | $\varnothing$ | $F$ | $E$ | $C$ |
| $E$ | $E$ | $C$ | $U$ | $A$ | $F$ | $\varnothing$ | $D$ | $B$ |
| $F$ | $F$ | $U$ | $C$ | $B$ | $E$ | $D$ | $\varnothing$ | $A$ |
| $U$ | $U$ | $F$ | $E$ | $D$ | $C$ | $B$ | $A$ | $\varnothing$ |

One difference between set algebra and ordinary algebra is immediately apparent. Whereas there are infinitely many numbers, there may be only finitely many sets in a set algebra.

*Example 4.1.* Let $U = \{a, b, c\}$. The subsets of $U$ are then eight in number, namely: $\varnothing$, $A = \{a\}$, $B = \{b\}$, $C = \{c\}$, $D = \{a, b\}$, $E = \{a, c\}$, $F = \{b, c\}$, and $U$. Tables 4.2 and 4.3 show the set algebra based on this $U$.

### Table 4.3

$$\alpha \cap \beta$$

| $\alpha \backslash \beta$ | $\varnothing$ | $A$ | $B$ | $C$ | $D$ | $E$ | $F$ | $U$ |
|---|---|---|---|---|---|---|---|---|
| $\varnothing$ | $\varnothing$ | $\varnothing$ | $\varnothing$ | $\varnothing$ | $\varnothing$ | $\varnothing$ | $\varnothing$ | $\varnothing$ |
| $A$ | $\varnothing$ | $A$ | $\varnothing$ | $\varnothing$ | $A$ | $A$ | $\varnothing$ | $A$ |
| $B$ | $\varnothing$ | $\varnothing$ | $B$ | $\varnothing$ | $B$ | $\varnothing$ | $B$ | $B$ |
| $C$ | $\varnothing$ | $\varnothing$ | $\varnothing$ | $C$ | $\varnothing$ | $C$ | $C$ | $C$ |
| $D$ | $\varnothing$ | $A$ | $B$ | $\varnothing$ | $D$ | $A$ | $B$ | $D$ |
| $E$ | $\varnothing$ | $A$ | $\varnothing$ | $C$ | $A$ | $E$ | $C$ | $E$ |
| $F$ | $\varnothing$ | $\varnothing$ | $B$ | $C$ | $B$ | $C$ | $F$ | $F$ |
| $U$ | $\varnothing$ | $A$ | $B$ | $C$ | $D$ | $E$ | $F$ | $U$ |

Notice that the operation tables given in this example are complete in the sense that every element of the algebra is represented in the tables. This contrasts with the usual addition and multiplication tables of arithmetic. An addition table, for example, gives the sum $x + y$ whenever $x$ and $y$ range from 0 to 9. Not only are numbers outside of the range of the table generated ($7 + 5 = 12$), but also we must have a set of rules governing the use of the tables for the addition of larger numbers such as 12 and 23.

We may then inquire as to the number of objects in a set algebra. Since these objects are subsets of a given (universal) set, the question is really that of how many subsets a given set has. Obviously a set with infinitely many elements has an infinite number of subsets. For example, the set of all positive integers has as subsets the sets $\{1\}$, $\{2\}$, $\{3\}$, $\{4\}$, . . ., as well as a great many others. The exact characterization of the number of subsets of an infinite set leads to the subject of transfinite cardinal numbers, and we shall not explore it further.*

The determination of the number of subsets of a finite set is simpler. If $A = \{x_1, \ldots, x_n\}$ is a set of $n$ elements and $B \subseteq A$, then each element $x_i \in A$ is either in $B$ or it is not in $B$. If we define $y_i$ to be 1 if $x_i \in B$ and 0 if $x_i \notin B$, then the set $B$ is characterized by a particular finite sequence of

* See, for example, Wilder [5].

zeros and ones, $y_1, \ldots, y_n$. Thus the number of subsets $B$ which can be formed is just the number of such sequences which can be written. Since there are two possibilities for each $y_i$ and $n$ $y_i$'s, there are $2^n$ such sequences. Hence we have the following result.

**Theorem 4.1.** A set of exactly $n$ elements has exactly $2^n$ subsets.

*Example 4.2.* Let $A = \{a, b, c\}$, and let $y_1$, $y_2$, and $y_3$ correspond to $a$, $b$, and $c$ respectively. The sequences and their corresponding subsets are given in Table 4.4.

**Table 4.4**

The sequences and sets of Example 4.2

| $y_1$ | $y_2$ | $y_3$ | Subset |
|:---:|:---:|:---:|:---:|
| 0 | 0 | 0 | $\varnothing$ |
| 0 | 0 | 1 | $\{c\}$ |
| 0 | 1 | 0 | $\{b\}$ |
| 0 | 1 | 1 | $\{b, c\}$ |
| 1 | 0 | 0 | $\{a\}$ |
| 1 | 0 | 1 | $\{a, c\}$ |
| 1 | 1 | 0 | $\{a, b\}$ |
| 1 | 1 | 1 | $A$ |

Two features of set algebra which are indicated in Tables 4.1 and 4.2 should be mentioned. First, the "unique number $-a$" specified in A4 becomes, for set algebra, the set $A$ itself. That is, $A \triangle A = \varnothing$ for any set $A$. Second, whereas for numbers it is generally not true that $a \cdot a = a$, for sets it is always true that $A \cap A = A$ for any set $A$. We express this property by saying that $A$ is an *idempotent*.

EXERCISES

1. The set algebra which we have developed is based on symmetric difference ($\triangle$) and intersection ($\cap$). Write the other set operations (union, relative complement, and complement) in terms of these two operations.
2. Express all of the set operations in terms of union and complement.
3. (a) Determine all subsets of the set $A = \{0, a, \#, 2\}$.
   (b) How many subsets does a set of ten elements have?
4. Write out the set algebras for the sets: (a) $U_1 = \{\mathbf{T}, \mathbf{F}\}$ and (b) $U_2 = \{0, a, \#, 2\}$.

## 5. PRODUCT SETS, RELATIONS, MAPPINGS

It is often useful to form pairs of related objects, as for example when pairing off partners at a dance, or when pairing off customers with suppliers. We may also on occasion relate more than two "objects," as when we list the distinguishing attributes of a botanical species. Mathematically, we speak of forming the product set.

**Definition 5.1.** Let $A$ and $B$ be two sets. The *product set $A \times B$* (read "$A$ cross $B$") is defined as $A \times B = \{\langle a, b \rangle \mid a \in A$ and $b \in B\}$. The symbol $\langle a, b \rangle$ is called an *ordered pair*, where the term "ordered" implies a distinction between the first and second element. If $A_1, A_2, \ldots, A_n$ are sets, then the *product set* is defined as $A_1 \times A_2 \times \cdots \times A_n = \{\langle a_1, a_2, \ldots, a_n \rangle \mid a_1 \in A_1, a_2 \in A_2, \ldots, a_n \in A_n\}$. The symbol $\langle a_1, a_2, \ldots, a_n \rangle$ is called an *ordered n-tuple*.

It should be noted that the product operation is not commutative. That is, $A \times B \neq B \times A$ in general.

*Example 5.1.* Let $A = \{2, 3\}$, $B = \{a, b, c\}$.

$$A \times B = \{\langle 2, a \rangle, \langle 2, b \rangle, \langle 2, c \rangle, \langle 3, a \rangle, \langle 3, b \rangle, \langle 3, c \rangle\}$$
$$B \times A = \{\langle a, 2 \rangle, \langle a, 3 \rangle, \langle b, 2 \rangle, \langle b, 3 \rangle, \langle c, 2 \rangle, \langle c, 3 \rangle\}$$
$$A \times A = \{\langle 2, 2 \rangle, \langle 2, 3 \rangle, \langle 3, 2 \rangle, \langle 3, 3 \rangle\}.$$

Frequently we are not interested in an entire product set, but only in a certain portion of it, which is in some way well defined. For example, if our product set consists of $n$-tuples of attributes, we are generally interested only in those combinations of attributes which actually occur. We usually have in mind some relationship defining the subset in which we are interested.

**Definition 5.2.** A *relation* is a subset of a product set. In particular, an *n-ary* relation is a subset of a product set on $n$ sets, $A_1, A_2, \ldots, A_n$.

*Example 5.2.* Let $A = \{1, 2, 3, 4, 5\}$. Then $A \times A$ has 25 elements. The relation *less than* ($a$ is less than $b$) is then $L = \{\langle 1, 2 \rangle, \langle 1, 3 \rangle, \langle 1, 4 \rangle, \langle 1, 5 \rangle, \langle 2, 3 \rangle, \langle 2, 4 \rangle, \langle 2, 5 \rangle, \langle 3, 4 \rangle, \langle 3, 5 \rangle, \langle 4, 5 \rangle\}$. The relation *equal* ($a$ is equal to $b$) is the set $E = \{\langle 1, 1 \rangle, \langle 2, 2 \rangle, \langle 3, 3 \rangle, \langle 4, 4 \rangle, \langle 5, 5 \rangle\}$

*Example 5.3.* Let $N$ be the set of non-negative integers, $W$ the set of English words. Let the relation $V \subseteq N \times W$ be defined as $V = \{\langle n, w \rangle \mid$ the word $w$ contains exactly $n$ occurrences of vowels$\}$. Then $V$ contains such ordered pairs as $\langle 1, a \rangle$, $\langle 1, the \rangle$, $\langle 3, sounded \rangle$, $\langle 5, facetious \rangle$, $\langle 2, been \rangle$, etc. Note that we could similarly define a relation $E \subseteq W \times N$ by $E = \{\langle w, n \rangle \mid$ the word $w$ contains exactly $n$ occurrences of vowels$\}$.

The last example illustrates a special class of relations which we shall need to use frequently. If we think of forming an ordered pair by selecting the first element of the pair first, and then the second element, we see that there is a vast difference between the relations $V$ and $E$ of the example.

For if $\langle a, b \rangle \in N \times W$ and we select a value of $a$ (that is, a positive integer), then, provided that we have selected a sufficiently small value, there are many values of $b$ (that is, words) such that $\langle a, b \rangle \in V$: there are many words containing exactly two vowel occurrences. On the other hand, if $\langle a, b \rangle \in W \times N$ and we select a value of $a$ (that is, a word), then there is exactly one way (barring variant spellings) to select a value of $b$ (that is, a number) so that $\langle a, b \rangle \in E$.

**Definition 5.3.** A binary relation $R$ is a *function* or *mapping* if whenever $\langle a, b \rangle \in R$ and $\langle a, c \rangle \in R$, then $b = c$. An $n$-ary relation $R_n$ is a *function* or *mapping* if whenever $\langle a_1, \ldots, a_{n-1}, b \rangle \in R_n$ and $\langle a_1, \ldots, a_{n-1}, c \rangle \in R_n$, then $b = c$. In particular, if $R \subseteq A_1 \times A_2 \times \cdots \times A_{n-1} \times A_n$ is a mapping, it is a mapping of $A_1 \times A_2 \times \cdots \times A_{n-1}$ into $A_n$. The mapping $R$ is *onto* if for every $b \in A_n$ there exist elements $a_i \in A_i$, $i = 1, \ldots, n - 1$, such that $\langle a_1, \ldots, a_{n-1}, b \rangle \in R$.

Looking ahead, we shall have need of mappings into the set of positive integers (length of words, number of parentheses, etc.), into the set $\{T, F\}$ (interpreted as "truth" and "falsity"), and into a set of graphs (the "tree" of a formula).

**Definition 5.4.** If $R \subseteq A \times B$ is a binary relation, then its *inverse relation* is $R^{-1} \subseteq B \times A$, defined by $\langle a, b \rangle \in R^{-1}$ if and only if $\langle b, a \rangle \in R$.

Thus $E$ and $V$ in Example 5.3 are inverse relations to each other. Note that the inverse of a function need not be a function. Note also that $(R^{-1})^{-1} = R$.

**Definition 5.5.** If $R$ and $R^{-1}$ are both functions or mappings, then $R$ (and $R^{-1}$) is said to be *one-to-one*.

The term "one-to-one" arises from the fact that one-to-one mappings send any two distinct elements into two distinct elements. Such mappings are important for many purposes, such as information retrieval. If we have assigned numbers to books (a mapping from the set of books into the set of numbers) then in order to use the numbers to locate the books we should have assigned distinct numbers to distinct books—that is, the mapping should be one-to-one.

## EXERCISES

1. Let $A = \{1, 3, 5\}$ and $B = \{2, 4\}$. Form $A \times B$, $B \times A$, $A \times A$, and $B \times B$.

2. If $A = \{a, b, i, k, t, z\}$ find the relation $P \subseteq A \times A$ defined by $\langle \alpha, \beta \rangle \in P$ if and only if $\alpha$ precedes $\beta$ in the usual alphabetic order.

3. Let $A = \{$the, quick, brown, fox, jumps, over, lazy, dog$\}$, and let $B$ be the set of positive integers. Determine the relation $E \subseteq A \times B \times B$, defined by $\langle a, b, c \rangle \in E$ if and only if the number of vowels in word $a$ is $b$ and the number of consonants is $c$.

4. For each of the following relations, determine the inverse relation. Is the given relation a mapping, and if so, is it one-to-one?
    (a) {⟨father, son⟩, ⟨daughter, mother⟩, ⟨father, brother⟩, ⟨1, mother⟩, ⟨2, 11⟩, ⟨1, brother⟩, ⟨father, 2⟩}
    (b) {⟨$a$, 1⟩, ⟨$b$, 1⟩, ⟨$c$, 2⟩, ⟨$d$, 3⟩}
    (c) {⟨$a$, 1⟩, ⟨$b$, 2⟩, ⟨$b$, 3⟩, ⟨$c$, 4⟩}
    (d) {⟨$a$, 1⟩, ⟨$b$, 2⟩, ⟨$c$, 3⟩, ⟨$d$, 4⟩}
5. Are the following relations mappings?
    (a) {⟨1, 2, $a$⟩, ⟨2, 1, $b$⟩, ⟨1, 7, $a$⟩}
    (b) {⟨1, 2, $a$⟩, ⟨3, 1, $b$⟩, ⟨1, 2, $c$⟩}
    (c) {⟨1, 2, 7, 8⟩, ⟨2, 1, 8, 7⟩, ⟨8, 2, 1, 7⟩, ⟨1, 2, 8, 7⟩}

## 6. THE SUBSET RELATION

We are now in a position to examine in more detail the one relationship which we have introduced between sets: the subset relation. This is, in fact, a relation in the sense that we have defined. For if we let $U$ be the universal set, and $V$ the set of all subsets of $U$, then $A \subseteq B$ is a relation in $V \times V$.

This particular relation has much in common with the relation "less than or equal to," defined for numbers. Both are *reflexive:* $n \leq n$, $A \subseteq A$; *antisymmetric:* if $n \leq m$ and $m \leq n$ then $n = m$, if $A \subseteq B$ and $B \subseteq A$ then $A = B$; and *transitive:* if $n \leq m$ and $m \leq p$ then $n \leq p$, if $A \subseteq B$ and $B \subseteq C$ then $A \subseteq C$; where these properties hold for all numbers $n$, $m$, and $p$, and all sets $A$, $B$, and $C$. Any relation having these three properties we call a *partial order*. Thus the subset relation is a partial order on the set of all subsets of $U$.

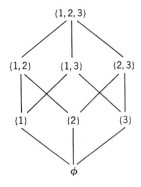

Figure 6.1. Lattice of subsets for {1, 2, 3}.

However, there are also differences between these two relations. Given any two numbers $m$ and $n$ either $m \leq n$ or $n \leq m$; but given the sets $A = \{1, 2\}$ and $B = \{2, 3\}$, for example, neither is a subset of the other. We may use a "Haase diagram" to indicate the subset relations within a given universal set. In such a diagram, $A \subseteq B$ if and only if $A$ lies below $B$ and is connected to it by a line. For example, Figure 6.1 is the diagram of subsets of the set $\{1, 2, 3\}$. Such a diagram is also known as a *lattice of subsets*.*

## EXERCISE

1. Construct the lattice of subsets for the set $\{a, b, c, d\}$.

## References

1. Arnold, B. H. *Logic and Boolean Algebra*. Prentice-Hall, Englewood Cliffs, New Jersey, 1962.
2. Rosenbloom, P. C. *The Elements of Mathematical Logic*. Dover Publications, New York, 1950.
3. Stoll, R. R. *Set Theory and Logic*. W. H. Freeman and Company, San Francisco, 1963.
4. Suppes, P. *Introduction to Logic*. D. Van Nostrand Company, Princeton, New Jersey, 1957.
5. Wilder, R. L. *Introduction to the Foundations of Mathematics*. John Wiley and Sons, New York, 1965. (Second Edition)

* For more information on lattices, see Arnold [1].

# 2. Boolean Algebras

## 1. DEFINITION

Boolean algebras, first studied in detail by George Boole [2], constitute an area of mathematics which has sprung into prominence with the advent of the digital computer. They are widely used in the design of switching circuits and computers, and are finding increasing applications in other areas. For our purposes they provide a link between set algebra and the propositional calculus.

**Definition 1.1.** A *Boolean algebra* is an algebraic system consisting of a set $S$ together with two operations $\#$ (as addition) and $\cdot$ (as multiplication) defined on the set, such that

1. the rules A1 to A5, M1 to M3, M5, D1, and D2 hold (see Chapter 1, Section 4),
2. every element is an idempotent, that is, if $a \in S$, then $a \cdot a = a$.

It is clear from this definition that every set algebra is a Boolean algebra, with $\triangle$ as addition and $\cap$ as multiplication. It is also true that every finite Boolean algebra is the algebra of all subsets of some universal set; but some of the infinite Boolean algebras form more complex set algebras.*

We have purposely avoided using "$+$" as the symbol for addition, since in most works on Boolean algebras this symbol is associated with the union ($\cup$) operation of the set algebras. We now define Boolean algebras in terms of this addition and multiplication. It can easily be shown that the two definitions are equivalent.

**Definition 1.2.** A *Boolean algebra* is an algebraic system consisting of a set $S$ together with two operations $+$ and $\cdot$ defined on the set, such that for any elements $a$, $b$, and $c$ of $S$ the following properties or *axioms* hold:

**A1.** $a + b \in S$
**M1.** $a \cdot b \in S$
**A2.** $a + (b + c) = (a + b) + c$
**M2.** $a \cdot (b \cdot c) = (a \cdot b) \cdot c$
**A3.** There exists a unique element $0 \in S$ such that for each $a \in S$, $a + 0 = 0 + a = a$.

---

* See Rosenbloom [8], Chapter 1.

**M3.** There exists a unique element $1 \in S$ such that for each $a \in S$, $a \cdot 1 = 1 \cdot a = a$.

**A5.** $a + b = b + a$

**M5.** $a \cdot b = b \cdot a$

**D1.** $a \cdot (b + c) = a \cdot b + a \cdot c$

**D2.** $(a + b) \cdot c = a \cdot c + b \cdot c$

**D3.** $a + (b \cdot c) = (a + b) \cdot (a + c)$

**D4.** $(a \cdot b) + c = (a + c) \cdot (b + c)$

**C1.** For each $a \in S$ there exists a unique element $a' \in S$ such that $a + a' = 1$ and $a \cdot a' = 0$.

With this definition, different properties of a Boolean algebra appear. There is no longer an additive inverse (A4), although we can still obtain one for a properly defined symmetric difference. Instead, we have axiom C1 with the $a'$ playing a role opposite from that of the usual inverse. In a set-algebraic interpretation, the $a'$ is just the complement of the set $a$. Here also we have four distributive axioms in place of the usual two (see Exercises Chapter 1, Section 3). We will now establish several more properties which Boolean algebras have.

EXERCISES

1. Define $\#$ in terms of $+$, $'$, and $\cdot$, and show that it has the property that for any $a \in S$ there is an element $(-a) \in S$ such that $a \# (-a) = (-a) \# a = 0$.
2. Construct a four-element Boolean algebra, giving the operation tables for the three operations $\#$, $+$, and $\cdot$.

## 2. DUALITY AND OTHER BASIC PROPERTIES

We will work entirely with the second definition that we have given for a Boolean algebra. We notice first of all that with the exception of A3, M3, and C1, the axioms given occur in pairs with the property that if the addition and multiplication are interchanged the only effect is to interchange the members of the pair—no new formulas are created. For example, if addition and multiplication are interchanged in D1, we obtain D3; and a similar interchange in D3 yields D1.

Such an interchange in A3, M3, and C1 yields new formulas. However, if we also interchange the 0 and 1, then the net effect is to transform A3 into M3 and vice versa, and to interchange the two formulas of C1. Thus we have an important *duality* in Boolean algebras.

### Principle of Duality

If a given formula is deducible from the axioms of a Boolean algebra, then the formula obtained by interchanging addition and multiplication, and the elements 0 and 1 throughout the given formula is also deducible.

This principle follows from the comments on the duality of the axioms, and the fact that each proof or deduction consists of a sequence of formulas which are axioms or are derived from the axioms. In the following chapter we will investigate more thoroughly the nature of a proof. For the moment, we present informal proofs of a few basic properties of Boolean algebras. In each of these we are assuming a Boolean algebra $\mathbf{B} = \langle S, +, \cdot \rangle$, and that all elements used are members of $S$. In addition, we shall drop the multiplication sign and write $ab$ in place of $a \cdot b$.

**Theorem 2.1.** For every element $a$, $a + a = a$ and $aa = a$.

*Proof.*
$$
\begin{aligned}
a + a &= (a + a)(1) & \text{(M3)} \\
&= (a + a)(a + a') & \text{(C1)} \\
&= a + aa' & \text{(D3)} \\
&= a + 0 & \text{(C1)} \\
&= a & \text{(A3)}
\end{aligned}
$$

The proof for $aa$ is obtained by duality:
$$
\begin{aligned}
aa &= aa + 0 & \text{(A3)} \\
&= aa + aa' & \text{(C1)} \\
&= a(a + a') & \text{(D1)} \\
&= a1 & \text{(C1)} \\
&= a & \text{(M3)}
\end{aligned}
$$

**Theorem 2.2.** For each element $a$, $a + 1 = 1$ and $a0 = 0$.

*Proof.*
$$
\begin{aligned}
a + 1 &= 1(a + 1) & \text{(M3)} \\
&= (a + a')(a + 1) & \text{(C1)} \\
&= a + a'1 & \text{(D3)} \\
&= a + a' & \text{(M3)} \\
&= 1 & \text{(C1)}
\end{aligned}
$$

The proof for $a0$ follows by duality.

**Theorem 2.3.** For each element $a$ and each element $b$, $a + ab = a$ and $a(a + b) = a$.

*Proof.*
$$
\begin{aligned}
a + ab &= a1 + ab & \text{(M3)} \\
&= a(1 + b) & \text{(D1)} \\
&= a1 & \text{(A5, Theorem 2.2)} \\
&= a & \text{(M3)}
\end{aligned}
$$

The proof for $a(a + b)$ follows by duality.

The formulas given in Theorem 2.3 are known as the *laws of absorption*, and they play a vital role in the simplification of Boolean formulas. It will be instructive for the reader to interpret these laws in a set algebra. We note also that they show clearly the relationships between the various distributive laws. For example,

$$
\begin{aligned}
(a + b)(a + c) &= aa + ac + ba + bc & \text{(D1, D2)} \\
&= a + ac + ab + bc & \text{(Theorem 2.1, M5)} \\
&= a + bc & \text{(Theorem 2.3, twice)}
\end{aligned}
$$

**Theorem 2.4.** For every $a$ and $b$, $(ab)' = a' + b'$ and $(a + b)' = a'b'$.

*Proof.* We know that $(ab)(ab)' = 0$ and $(ab) + (ab)' = 1$, and that for $ab$, $(ab)'$ is the unique element for which these statements hold, by C1. Thus if we can show that $a' + b'$ in place of $(ab)'$ satisfies these same equations, we have the first formula of the theorem. The second follows by duality. Hence we look at $(ab)(a' + b')$ and $ab + a' + b'$.

$$
\begin{aligned}
(ab)(a' + b') &= aba' + abb' & \text{(D1, M2)} \\
&= 0b + a0 & \text{(M2, M5, C1)} \\
&= 0 + 0 & \text{(Theorem 2.2)} \\
&= 0 & \text{(A3)} \\
ab + a' + b' &= (a + a' + b')(b + a' + b') & \text{(D4)} \\
&= (1 + b')(1 + a') & \text{(A2, A5, C1)} \\
&= 1 \cdot 1 & \text{(Theorem 2.2)} \\
&= 1 & \text{(M3)}
\end{aligned}
$$

Thus it follows that $(ab)' = a' + b'$.

The formulas in Theorem 2.4 are known as *DeMorgan's Laws*, after Augustus DeMorgan.

**Theorem 2.5.** $0' = 1$ and $1' = 0$.

*Proof.* By Theorem 2.2, $1 + 0 = 1$ and $1 \cdot 0 = 0$. But these are the equations of C1, and by the uniqueness postulated there, $1' = 0$ and $0' = 1$.

**Theorem 2.6.** If a Boolean algebra contains at least two distinct elements, then $0 \neq 1$.

*Proof.* Suppose that there is a Boolean algebra with at least two distinct elements, for which $0 = 1$. Let $a$ be an element distinct from 0. We know such an element exists, for otherwise *all* elements are equal to 0. But then $a = a1 = a0 = 0$, which is a contradiction. Hence $0 \neq 1$.

Since every set algebra is a Boolean algebra, the partial order which is the subset relation must hold for at least some Boolean algebras. As a matter of fact, it holds for all of them.

**Definition 2.1.** Let $x$ and $y$ be elements of a Boolean algebra. We say that *x is less than or equal to y* $(x \leq y)$ if and only if $x + y = y$. (See Exercise 5 of Chapter 1, Section 3.)

**Theorem 2.7.** $\leq$ is a partial order.

*Proof.* By Theorem 2.1 $x + x = x$. Thus $x \leq x$. If $x \leq y$ then $x + y = y$; if $y \leq x$ then $x + y = y + x = x$. Thus if $x \leq y$ and $y \leq x$, $x = y$. Finally, suppose that $x \leq y$ and $y \leq z$. Then $x + y = y$ and $y + z = z$. Hence $x + z = x + (y + z) = (x + y) + z = y + z = z$. Thus $x \leq z$.

**Theorem 2.8.** Let $x$, $y$, and $z$ be elements of a Boolean algebra. Then the partial order $\leq$ has the following properties.

(i) If $x \leq y$ and $x \leq z$, then $x \leq yz$.

(ii) If $x \leq y$, then $x \leq y + z$ for any element $z$.

(iii) If $x \leq y$, then $xz \leq y$ for any element $z$.

(iv) $x \leq y$ if and only if $y' \leq x'$.

*Proof.* (i) $x + y = y$ and $x + z = z$. Thus $x + yz = (x + y)(x + z)$
$= yz$.

(ii) If $x + y = y$, then $x + (y + z) = (x + y) + z = y + z$.

(iii) By the absorption laws, $xz + x = x$, or $xz \leq x$. The result then follows by transitivity.

(iv) Suppose $x \leq y$. Then $x + y = y$ and hence $y' = (x + y)'$. Thus $y' + x' = (x + y)' + x' = ((x + y)x)' = x'$, by the absorption laws. The converse follows since $(x')' = x$ by Exercise 2 of this section.

EXERCISES

1. Prove that for any elements $a$ and $b$ of a Boolean algebra, $a + a'b = a + b$ and $a(a' + b) = ab$.
2. Prove that for any element $a$ of a Boolean algebra, $(a')' = a$.
3. Prove that for any elements $a$ and $b$ of a Boolean algebra, $a \leq b$ if and only if $ab' = 0$.
4. Prove that for any elements $a$ and $b$ of a Boolean algebra, $a \leq b$ if and only if $b + a' = 1$.
5. Show that no Boolean algebra has exactly three elements.
6. Show that no finite Boolean algebra with more than one element has an odd number of elements.

## 3. BOOLEAN FUNCTIONS

In this section we wish to discuss Boolean functions, that is, mappings (or functions) from a Boolean algebra into itself. We shall develop a standard form for these functions and discuss briefly various representations of them. The type of definition which we shall use is that known as a "recursive" definition. In it two simple types of Boolean functions are defined, and rules are given for constructing all other Boolean functions from these.

**Definition 3.1.** Let $x_1, \ldots, x_n$ be variables whose values lie in a given Boolean algebra. A mapping $f$ of the Boolean algebra into itself is a *Boolean function of n variables* if it can be constructed according to the following rules.

1. If for any values of $x_1, \ldots, x_n, f(x_1, \ldots, x_n) = a$, where $a$ is a fixed element of the Boolean algebra, then $f$ is a Boolean function. This is the *constant* function.

2. If for any values of $x_1, \ldots, x_n, f(x_1, \ldots, x_n) = x_i$ for some $i = 1, \ldots, n$, then $f$ is a Boolean function. This is a *projection* function.

3. If $f$ is a Boolean function, then $g$, defined by $g(x_1, \ldots, x_n) = (f(x_1, \ldots, x_n))'$ for all $x_1, \ldots, x_n$, is a Boolean function.

4. If $f$ and $g$ are Boolean functions, then $h$ and $k$, defined by

$$h(x_1, \ldots, x_n) = f(x_1, \ldots, x_n) + g(x_1, \ldots, x_n)$$

and

$$k(x_1, \ldots, x_n) = f(x_1, \ldots, x_n)g(x_1, \ldots, x_n)$$

for all $x_1, \ldots, x_n$, are Boolean functions.

5. Any function which can be constructed by a finite number of applications of the above rules and only such a function is a Boolean function.

Thus a Boolean function is any function which can be constructed from the constant and projection functions by finitely many uses of the operations $'$, $+$, and $\cdot$. For a function of one variable, the projection function is the *identity* function: $f(x) = x$.

*Example 3.1.* The following are Boolean functions, where the variables $x$, $y$, and $z$ range over a Boolean algebra and $a$ is an element of this algebra.

$$f(x) = x + x'a$$
$$g(x, y) = x'y + xy' + y'$$
$$h(x, y, z) = axy'z + yz' + a + xy$$

Because of the relationships which hold between the operations, it is possible for a Boolean function to take on many forms. For example, if $f(x, y) = x'y'$ and $g(x, y) = (x + y)'$, then by DeMorgan's Laws we know that $f$ and $g$ are the same function—that is, they take on identical values for identical values of the variables. Thus, in order to determine better whether or not two expressions represent the same Boolean function, it is desirable to have a standard or *canonical form* into which the expressions can be transformed. We develop such a form in this theorem and the comments which follow it.

**Theorem 3.1.** If $f$ is a Boolean function of one variable, then for all values of $x, f(x) = f(1)x + f(0)x'$.

*Proof.* We examine the possible forms of $f$.

*Case 1.* $f$ is a constant function, say $f(x) = a$.

$$f(1)x + f(0)x' = ax + ax' = a(x + x') = a1 = a = f(x)$$

*Case 2.* $f$ is the identity function.

$$f(1)x + f(0)x' = 1x + 0x' = x + 0 = x = f(x)$$

*Case 3.* Suppose the theorem holds for $f$ and let $g(x) = (f(x))'$.

$$\begin{aligned}
g(x) = (f(x))' &= (f(1)x + f(0)x')' \\
&= (f(1)x)'(f(0)x')' \\
&= ((f(1))' + x')((f(0))' + x) \\
&= (f(1))'(f(0))' + (f(1))'x + (f(0))'x' + xx' \\
&= (f(1))'(f(0))'(1) + (f(1))'x + (f(0))'x' \\
&= (f(1))'(f(0))'(x + x') + (f(1))'x + (f(0))'x' \\
&= (f(1))'(f(0))'x + (f(1))'x \\
&\qquad\qquad + (f(1))'(f(0))'x' + (f(0))'x' \\
&= (f(1))'x + (f(0))'x' \quad \text{(by absorption)} \\
&= g(1)x + g(0)x'
\end{aligned}$$

*Case 4.* Suppose the theorem holds for $f$ and $g$, and let $h(x) = f(x) + g(x)$.

$$\begin{aligned}
h(x) &= f(x) + g(x) \\
&= f(1)x + f(0)x' + g(1)x + g(0)x' \\
&= (f(1) + g(1))x + (f(0) + g(0))x' \\
&= h(1)x + h(0)x'
\end{aligned}$$

*Case 5.* Suppose the theorem holds for $f$ and $g$, and let $k(x) = f(x)g(x)$.

$$\begin{aligned}
k(x) &= f(x)g(x) \\
&= (f(1)x + f(0)x')(g(1)x + g(0)x') \\
&= f(1)g(1)xx + f(1)g(0)xx' + f(0)g(1)x'x + f(0)g(0)x'x' \\
&= f(1)g(1)x + f(0)g(0)x' \\
&= k(1)x + k(0)x'
\end{aligned}$$

Thus we have established a canonical form for a Boolean function of one variable. In a similar manner we may show that if $f$ is a Boolean function of two variables, then for all values of $x$ and $y$

$$f(x, y) = f(1, 1)xy + f(1, 0)xy' + f(0, 1)x'y + f(0, 0)x'y'$$

And in general, if $f$ is a Boolean function of $n$ variables, then for all values of $x_1, \ldots, x_n$

$$f(x_1, \ldots, x_n) = \sum f(e_1, \ldots, e_n)x_1^{e_1}x_2^{e_2} \cdots x_n^{e_n}$$

where the $e_i$ take on the values 0 and 1, and $x_i^{e_i}$ is interpreted as $x_i$ or $x_i'$ accordingly as $e_i$ has the value 1 or 0.

*Example 3.2.* Let $B$ be a four-element Boolean algebra with elements 0, $a$, $a'$, and 1. We construct the canonical forms of the first two functions of Example 3.1. The values of these functions are given in Tables 3.1 and 3.2 respectively.

Now $f(1) = 1$ and $f(0) = a$, so that the canonical form for $f$ is $f(1)x + f(0)x' = 1x + ax'$. Note that this is easily reducible to the original form of $f(x)$. Similarly, $g(1, 1) = 0$ and $g(1, 0) = g(0, 1) = g(0, 0) = 1$, so that the canonical form for $g$ is $g(x, y) = 0xy + 1xy' + 1x'y + 1x'y'$, which is again easily reducible to the given form.

Table 3.1

Values of $f(x) = x + x'a$

| $x$ | $f(x)$ |
| --- | --- |
| 0 | $a$ |
| $a$ | $a$ |
| $a'$ | 1 |
| 1 | 1 |

Table 3.2

Values of $g(x, y) = x'y + xy' + y'$

| $x\backslash y$ | 0 | $a$ | $a'$ | 1 |
| --- | --- | --- | --- | --- |
| 0 | 1 | 1 | 1 | 1 |
| $a$ | 1 | $a'$ | 1 | $a'$ |
| $a'$ | 1 | 1 | $a$ | $a$ |
| 1 | 1 | $a'$ | $a$ | 0 |

The canonical form which we have been discussing is known as a *sum of products* or *disjunctive normal form*. There is also a *product of sums* or *conjunctive normal form*. These are also sometimes referred to as *minterm* and *maxterm* normal forms, respectively. We shall study them in more detail in Chapter 3.

It is evident from the disjunctive normal form that for any Boolean algebra, the values of any Boolean function $f$ of $n$ variables are completely determined by its values at the points $\langle x_1, \ldots, x_n \rangle$ where each of the $x_i$ has the value 0 or 1. This provides a nice geometric interpretation of Boolean functions which is exploited, for example, by Miller [5], and Flegg [3].

EXERCISES

1. Suppose that $f$ is a Boolean function of 1 variable on a four-element Boolean algebra, $f(0) = a'$ and $f(1) = a$. Determine an expression for $f$.
2. Write out the general canonical form for a Boolean function of three variables.

3. Determine the canonical form for each of the following functions.
   (a) $f(x) = xx'$
   (b) $f(x, y) = xy' + ax' + by$, where $a$ and $b$ are distinct fixed elements of the Boolean algebra.
   (c) $f(x, y, z) = x(y + az') + (x' + z)(ax + y' + z)$
4. Suppose that $B$ is a Boolean algebra on the set $\{0, a, a', b, b', c, c', 1\}$, and let $f$ be a Boolean function such that $f(0, 0, 0) = f(0, 0, 1) = f(1, 0, 0) = a, f(0, 1, 0) = 0, f(0, 1, 1) = 1, f(1, 0, 1) = f(1, 1, 0) = c'$, and $f(1, 1, 1) = b$. Determine $f(a', c, b)$.

## 4. MINIMIZATION OF BOOLEAN FUNCTIONS

The canonical forms of Boolean functions discussed in the last section are useful for several purposes such as determining whether or not two expressions represent the same function. But for other purposes they are often unwieldy in the sense that they involve more operations than are necessary. For example, the two functions used in Example 3.2 can be equally well represented by the expressions $f(x) = x + a$, and $g(x, y) = x' + y'$ or $g(x, y) = (xy)'$. Particularly when hardware is being constructed to implement Boolean functions, the question of determining a minimal expression for a function is often crucial. In this section we wish to examine this problem and give some techniques for handling it.

The first problem in minimizing a given function, that is, finding the "simplest" expression representing this function, is that of defining exactly what the criteria are for determining how simple an expression is. Arriving at this definition is not always simple: for example, the two expressions $xy'z + x'y + x'z'$ and $(x + y + z')(x' + y')(x' + z)$ represent the same Boolean function, and who can say which is the simpler? They both contain the same number of occurrences of each variable and the same number of operations. The problem becomes more complicated when one is willing to allow other basic operations, as is often the case when hardware is being designed to implement some function. For example, if one is allowed to use the symmetric difference operation, this same function can be represented by $x \# (y + z')$. While this involves fewer individual operations, it may or may not be simpler to implement with a given type of equipment.

We shall assume here that "simplest form" means "simplest sum of products form," and shall discuss two techniques for achieving such a form. Since we are working in a Boolean algebra, one could in theory perform a simplification entirely algebraically, utilizing the absorption laws and similar relationships. The difficulty lies in being sure that one has taken into account all possible ways of applying the simplifying rules. The

Figure 4.1. One-variable Karnaugh map.    Figure 4.2. Two-variable Karnaugh map.

techniques which have been evolved to aid in simplification are then basically methods of arranging information about the function so that one does see all possibilities.

The first method which we shall discuss is graphical: the use of Karnaugh maps or diagrams. A Karnaugh diagram is in reality a Venn diagram with the various regions arranged as squares within a rectangle. As such, it suffers from the same defect which a Venn diagram has: for more than six variables, Karnaugh maps become too complicated to use; and even for five or six variables the maps lose much of their utility. The Karnaugh maps for one through four variables are presented in Figures 4.1 to 4.4. Each square represents the product of the coordinates given on the edges of the rectangles. For example, the square labeled "$a$" in Figure 4.3 represents the term $x'yz$, and that labeled "$b$" in Figure 4.4 represents the term $wx'y'z'$.

Conventionally, each square in a Karnaugh map contains either a "1" if the term represented by that square is required to be present in a sum of products representation of the function, or a "0" if the term is required to be absent, or a "$d$" ("don't care") if the presence or absence of the term is immaterial. Often the zeros are not explicitly written down. The method of using such a map is to attempt to cover all of the 1's with the least possible number of rectangles which represent products of elements. For example, in Figure 4.5 (where a simplified coordinate system is used) the two squares marked "$a$" represent the term $wxy'z' + wxyz'$, which can be simplified to $wxz'$ by use of the distributive laws. Similarly the squares marked "$b$" represent the term $w'xy'$, and the four squares marked "$c$" correspond to the term $yz$. However, the two squares marked "$d$" corre-

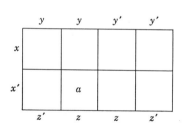

Figure 4.3. Three-variable Karnaugh map.

x x x' x'
w b y'
w y
w' y
w' y'
z' z z z'

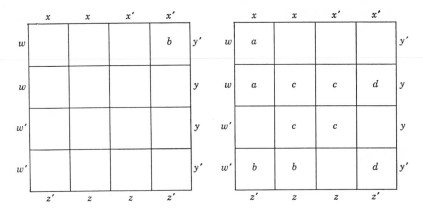

Figure 4.4. Four-variable Karnaugh map.

x x x' x'
w a y'
w a c c d y
w' c c y
w' b b d y'
z' z z z'

Figure 4.5. Examples of regions on a Karnaugh map.

spond to $wx'yz' + w'x'y'z'$, which cannot be further simplified. And in general a group of three squares for example has no significance. The meaningful groupings always consist of contiguous squares, provided that we regard the opposite edges of the map as contiguous. Thus, in Figure 4.6 the squares marked "*a*," "*b*," and "*c*" represent respectively the terms $x'y'z$, $yz'$, and $y'z'$. Within these regions each square must contain either a 1 or a *d*. No square may contain a 0, for these represent terms explicitly excluded; not all *d*'s must be covered as we don't care whether the corresponding terms are present.

*Example 4.1.* Let us consider the function whose Karnaugh map is

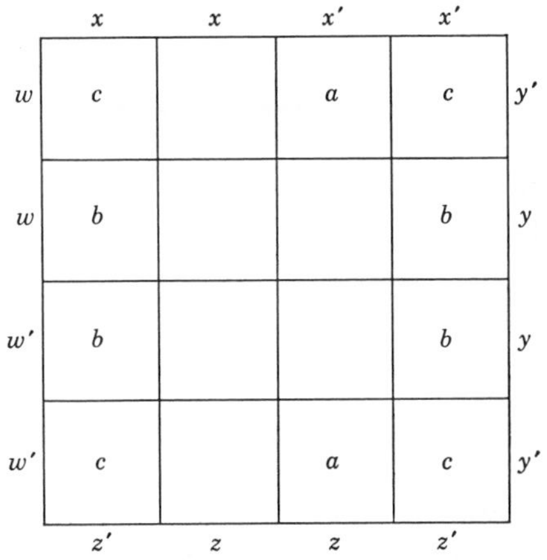

x x x' x'
w c a c y'
w b b y
w' b b y
w' c a c y'
z' z z z'

Figure 4.6. Further Karnaugh map regions.

given in Figure 4.7. We have drawn on this map three rectangles which cover the 1's and represent the terms $wxz'$, $xy$, and $w'x'z$. Notice that the square marked "$a$" could have been used by itself, but by combining it with the square below it we obtain the simpler term $wxz'$ in place of $wxy'z'$. The latter square is then included in two terms, but this does not matter since $x + x = x$ for any $x$. The two squares marked "$b$" represent the product $w'yz$, but it is not necessary to include this term as the squares are covered by other terms.

*Example 4.2.* We now examine a situation with "don't care" terms, as represented in Figure 4.8. Such terms may arise because in a particular application we know that certain products will never occur. We have shown here a covering of the 1's representing the function by the expression $xz + yz'$.

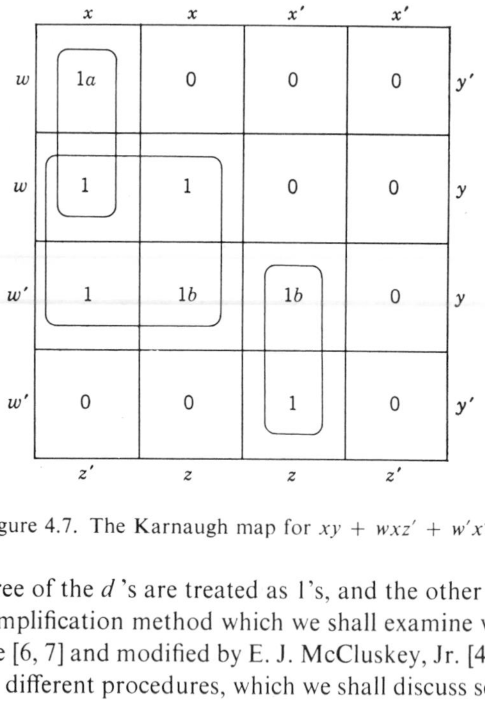

| | $x$ | $x$ | $x'$ | $x'$ | |
|---|---|---|---|---|---|
| $w$ | 1a | 0 | 0 | 0 | $y'$ |
| $w$ | 1 | 1 | 0 | 0 | $y$ |
| $w'$ | 1 | 1b | 1b | 0 | $y$ |
| $w'$ | 0 | 0 | 1 | 0 | $y'$ |
| | $z'$ | $z$ | $z$ | $z'$ | |

Figure 4.7. The Karnaugh map for $xy + wxz' + w'x'z$.

Notice that three of the $d$'s are treated as 1's, and the other two as 0's.

The other simplification method which we shall examine was originated by W. V. Quine [6, 7] and modified by E. J. McCluskey, Jr. [4]. The method consists of two different procedures, which we shall discuss separately. The first part of the method is a simplification based on an application of the distributive law: if $A$ is any Boolean expression and $x$ is a Boolean variable, then $Ax + Ax' = A(x + x') = A$. Thus if two products differ only in that *one* variable is complemented in one but not in the other, the two products may be combined into one by omitting that one variable. The Quine method consists of beginning with the full disjunctive normal form in which all variables appear in each product and forming all possible combinations of products by using this rule. The procedure is then iterated on all of the terms obtained, until we arrive at a set of products, known as the *prime implicants*, which can be combined no further. The given function may then be written as a sum of the prime implicants.

McCluskey's modification of this method lies in providing a numerical representation for the terms and in improved procedures for the second part. If an uncomplemented variable is represented by "1" and a complemented variable by "0", then, for example, the term $wx'y'z$ may be represented by the binary numeral 1001. If the numerical terms are then grouped by the number of ones, those terms which can be combined lie in groups

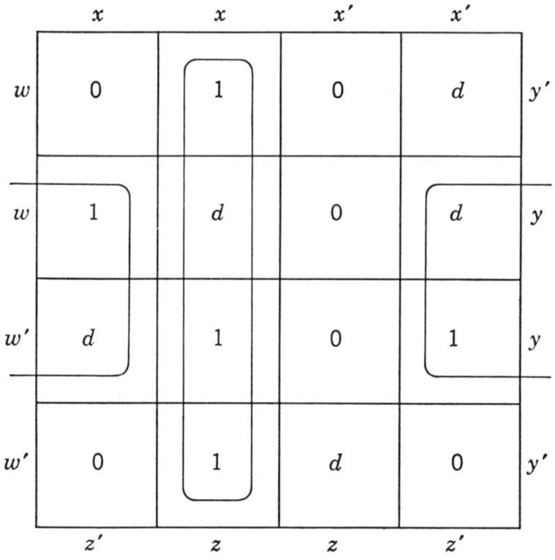

Figure 4.8. A Karnaugh map with "don't care" conditions.

whose numbers of ones differ by one. Thus the search for prime implicants is simplified.

*Example 4.3.* Referring to the Karnaugh map of Figure 4.7, we find that the full disjunctive normal form for this function is the expression $wxyz + wxyz' + wxy'z' + w'xyz + w'xyz' + w'x'yz + w'x'y'z$. The numerical representation of these terms, grouped according to the number of ones, is

$$
\begin{array}{ll}
w'x'y'z & \{0001 \\
w'x'yz & \lceil 0011 \\
w'xyz' & \{ 0110 \\
wxy'z' & \lfloor 1100 \\
w'xyz & \lceil 0111 \\
wxyz' & \{ 1110 \\
wxyz & \{1111\text{--}
\end{array}
$$

Comparing terms in adjacent groups, we find the following combinations, where the dash represents the dropped variable:

$$\{00\text{–}1*$$

$$\left\{\begin{array}{l}0\text{–}11* \\ 011\text{–} \\ \text{–}110 \\ 11\text{–}0* \end{array}\right.$$

$$\left\{\begin{array}{l}\text{–}111 \\ 111\text{–}\end{array}\right.$$

Again comparing adjacent groupings we find that those numerical expressions which are starred fail to combine with any others, and hence are prime implicants. The other terms combine to yield just one new term, –11–, which is, of course, a prime implicant since it has no other term with which to combine. The prime implicants are thus the four products $w'x'z$, $w'yz$, $wxz'$, and $xy$, and hence the given function may be represented by the expression $w'x'z + w'yz + wxz' + xy$.

As can be seen from this last example, although the Quine-McCluskey method yields a relatively simple expression for a given function at this point, it does not necessarily yield the simplest. In this particular example the solution by Karnaugh maps showed that the term $w'yz$ is not necessary. This term may be removed algebraically, using the absorption laws. In fact,

$$w'yz = w'1yz = w'(x' + x)yz = w'x'yz + w'xyz$$

Then the first of these terms may be absorbed by $w'x'z$ and the second by $xy$. Thus the term is removed by a procedure rather different from that used in the first part of the Quine-McCluskey method.

It should be remarked that the expansion to the full disjunctive normal form is vital for the Quine-McCluskey method. For example, without such expansion, the Quine-McCluskey method fails to reduce the expression $wx + x'y + wx'y'$. But after expansion, this expression reduces to $w + x'y$.

The second part of this method is a tabular procedure designed to eliminate the redundant prime implicants. A numeral is assigned to each product in the full disjunctive normal form of the functions: it is convenient to use the decimal numeral corresponding to the binary numeral assigned in the first part of this method. Similarly, each prime implicant is assigned a letter. A table is then formed showing terms in the full disjunctive normal form versus prime implicants. A cross ( × ) is placed in the $(j, i)$ position of the table if the $i$th term was involved in the formation of the $j$th prime implicant.

If the column of any term contains a single cross, then the prime implicant in whose row the × lies is essential to represent that particular term. Thus we mark that prime implicant and delete its row from the table. Furthermore, any term having a cross in that row is represented by that prime implicant, and hence its column may also be deleted from the table. If this procedure rules out all columns, then the given function is represented by the sum of the marked prime implicants. However, if additional columns remain, further prime implicants are marked, each time choosing one representing a maximal number of the remaining terms. Corresponding columns are deleted from the table, and this procedure is continued until there are no columns remaining. A simple expression for the given function is then the sum of the marked prime implicants.

*Example 4.4.* We continue the process for the function of Example 4.3. The decimal numerals corresponding to the original terms are 1, 3, 6, 7, 12, 14, and 15; and we may letter the prime implicants $A: w'x'z$, $B: w'yz$, $C: wxz'$, and $D: xy$. Table 4.1 then shows the contribution of each term

**Table 4.1**

Prime implicant chart for Example 4.4

|   | 1 | 3 | 6 | 7 | 12 | 14 | 15 |
|---|---|---|---|---|----|----|----|
| $A$ | ⊗ | × |   |   |    |    |    |
| $B$ |   | × |   | × |    |    |    |
| $C$ |   |   |   |   | ⊗  | ×  |    |
| $D$ |   |   | ⊗ | × |    | ×  | ×  |

to the prime implicants. The essential crosses are encircled and the columns deleted by their choice are: $A$–1 and 3; $C$–12 and 14; $D$–6, 7, (14), and 15. Thus in this example all columns are deleted at this stage and hence the function may be represented by the sum of $A$, $C$, and $D$, or $w'x'z + wxz' + xy$, the same result as obtained by use of the Karnaugh map.

*Example 4.5.* As our final example in this section we take the function

$$f(v, w, x, y, z) = vwxyz + vwxyz' + vwxy'z + vwx'yz' + vwx'y'z$$
$$+ vwx'y'z' + vw'xyz + vw'xyz' + vw'xy'z$$
$$+ vw'x'yz' + v'wxyz + v'wxy'z' + v'wx'yz$$
$$+ v'w'xy'z + v'w'xy'z' + v'w'x'yz' + v'w'x'y'z'$$

The first part of the Quine-McCluskey reduction for this function is presented in Table 4.2, and the prime implicant chart for the second part in Table 4.3. In this table the two essential crosses have been encircled. The deletion of rows and columns related to these crosses gives us the

## Table 4.2

Quine-McCluskey reduction, part one

| (a) | (b) | (c) | (d) | (e) | (f) | (g) | (h) | (i) |
|---|---|---|---|---|---|---|---|---|
| $v'w'x'y'z'$ | 0 | 00000 | 000–0 | ( 0, 2) | A | 1––10 | (18, 22, 26, 30) | L |
| | | | ––––– | 00–00 | ( 0, 4) | B | ––––– | |
| $v'w'x'yz'$ | 2 | 00010 | ––––– | | | 1–1–1 | (21, 23, 29, 31) | M |
| $v'w'xy'z'$ | 4 | 00100 | –0010 | ( 2, 18) | C | 1–11– | (22, 23, 30, 31) | N |
| | | | ––––– | 0010– | ( 4, 5) | D | | |
| $v'w'xy'z$ | 5 | 00101 | 0–100 | ( 4, 12) | E | | |
| $v'wxy'z'$ | 12 | 01100 | ––––– | | | | |
| $vw'x'yz'$ | 18 | 10010 | –0101 | ( 5, 21) | F | | |
| $vwx'y'z'$ | 24 | 11000 | 10–10 | (18, 22) | | | |
| | | | ––––– | 1–010 | (18, 26) | | | |
| $v'wx'yz$ | 11 | 01011 | 1100– | (24, 25) | G | | |
| $vw'xy'z$ | 21 | 10101 | 110–0 | (24, 26) | H | | |
| $vw'xyz'$ | 22 | 10110 | ––––– | | | | |
| $vwx'y'z$ | 25 | 11001 | 01–11 | (11, 15) | I | | |
| $vwx'yz'$ | 26 | 11010 | 101–1 | (21, 23) | | | |
| | | | ––––– | 1–101 | (21, 29) | | | |
| $v'wxyz$ | 15 | 01111 | 1011– | (22, 23) | | | |
| $vw'xyz$ | 23 | 10111 | 1–110 | (22, 30) | | | |
| $vwxy'z$ | 29 | 11101 | 11–01 | (25, 29) | J | | |
| $vwxyz'$ | 30 | 11110 | 11–10 | (26, 30) | | | |
| $vwxyz$ | 31 | 11111 | –1111 | (15, 31) | K | | |
| | | | 1–111 | (23, 31) | | | |
| | | | 111–1 | (29, 31) | | | |
| | | | 1111– | (30, 31) | | | |

(a) Term in full disjunctive normal form.
(b) Decimal numeral of term.
(c) Binary numeral of term.
(d) First reduction.
(e) Terms yielding configuration (d).
(f) Labels of prime implicants.
(g) Second reduction.
(h) Terms yielding configuration (g).
(i) Labels of prime implicants.

reduced prime implicant chart of Table 4.4. From this it is clear that the next prime implicant chosen must be $L$, $M$, or $N$, since each of these has four crosses in its row. If $L$ is chosen and the corresponding columns deleted, then $M$ still has four crosses while $N$ is reduced to two crosses.

**Table 4.3**

Prime implicant chart for Example 4.5

|   | 0 | 2 | 4 | 5 | 11 | 12 | 15 | 18 | 21 | 22 | 23 | 24 | 25 | 26 | 29 | 30 | 31 |
|---|---|---|---|---|----|----|----|----|----|----|----|----|----|----|----|----|----|
| A | × | × |   |   |    |    |    |    |    |    |    |    |    |    |    |    |    |
| B | × |   | × |   |    |    |    |    |    |    |    |    |    |    |    |    |    |
| C |   | × |   |   |    |    |    | ×  |    |    |    |    |    |    |    |    |    |
| D |   |   | × | × |    |    |    |    |    |    |    |    |    |    |    |    |    |
| E |   |   | × |   |    | ⊗  |    |    |    |    |    |    |    |    |    |    |    |
| F |   |   |   | × |    |    |    |    | ×  |    |    |    |    |    |    |    |    |
| G |   |   |   |   |    |    |    |    |    |    |    | ×  | ×  |    |    |    |    |
| H |   |   |   |   |    |    |    |    |    |    |    | ×  |    | ×  |    |    |    |
| I |   |   |   |   | ⊗  |    | ×  |    |    |    |    |    |    |    |    |    |    |
| J |   |   |   |   |    |    |    |    |    |    |    |    | ×  |    | ×  |    |    |
| K |   |   |   |   |    |    | ×  |    |    |    |    |    |    |    |    |    | ×  |
| L |   |   |   |   |    |    |    | ×  |    | ×  |    |    |    | ×  |    | ×  |    |
| M |   |   |   |   |    |    |    |    | ×  |    | ×  |    |    | ×  |    |    | ×  |
| N |   |   |   |   |    |    |    |    |    | ×  | ×  |    |    |    |    | ×  | ×  |

**Table 4.4**

Reduced prime implicant chart

|   | 0 | 2 | 5 | 18 | 21 | 22 | 23 | 24 | 25 | 26 | 29 | 30 | 31 |
|---|---|---|---|----|----|----|----|----|----|----|----|----|----|
| A | × | × |   |    |    |    |    |    |    |    |    |    |    |
| B | × |   |   |    |    |    |    |    |    |    |    |    |    |
| C |   | × |   | ×  |    |    |    |    |    |    |    |    |    |
| D |   |   | × |    |    |    |    |    |    |    |    |    |    |
| F |   |   | × | ×  |    |    |    |    |    |    |    |    |    |
| G |   |   |   |    |    |    |    | ×  | ×  |    |    |    |    |
| H |   |   |   |    |    |    | ×  |    | ×  |    |    |    |    |
| J |   |   |   |    |    |    |    | ×  |    | ×  |    |    |    |
| K |   |   |   |    |    |    |    |    |    |    |    |    | ×  |
| L |   |   |   | ×  |    | ×  |    |    |    | ×  | ×  |    |    |
| M |   |   |   |    | ×  |    | ×  |    |    | ×  |    | ×  |    |
| N |   |   |   |    |    | ×  | ×  |    |    |    | ×  | ×  |    |

Hence the next choice would be $M$. Similarly, if $M$ were chosen first the next choice would be $L$. However, if $N$ were chosen first, then both $L$ and $M$ would be reduced to two crosses, so that the next choice would be from the set $\{A, C, F, G, H, J, L, M\}$. Thus at this stage the Quine-McCluskey method as we have described it becomes indeterminate. For this particular example the following sums of prime implicants, among others, can be derived as representing the function:

$$A + C + D + E + G + H + I + M + N$$
$$A + D + E + G + I + L + M$$
$$A + E + F + G + I + J + L + N$$
$$A + E + F + G + I + L + M$$
$$A + E + F + G + I + L + M + N$$
$$A + E + F + H + I + J + L + M + N$$

The indeterminacy shown in this last example can be reduced somewhat by being a little sophisticated. Since we are seeking a simple representation for the function our choices of prime implicants should be guided not only by the number of terms which they represent but also by some measure of simplicity. For example, one might assign to each prime implicant an ordered pair $\langle a, b \rangle$ where $a$ is the total number of variables occurring in the prime implicant, and $b$ is the number of complemented variables.

**Definition 4.1.** Let $A$ and $B$ be Boolean products with associated ordered pairs $\langle a, b \rangle$ and $\langle c, d \rangle$ respectively. We say that $A$ *is simpler than* $B$ if either $a < c$, or $a = c$ and $b < d$.

This relation is then a partial order on the prime implicants of a given formula and helps to reduce the indeterminacy, although it does not eliminate it. For example, in the function of Example 4.5 the prime implicants $L$, $M$, and $N$ have associated pairs $\langle 3, 1 \rangle$, $\langle 3, 0 \rangle$, and $\langle 3, 0 \rangle$ respectively. Thus either $M$ or $N$ is simpler than $L$, but one still has a choice between $M$ and $N$. Finally, one can define an ordered pair for a sum of prime implicants by summing each component of the ordered pairs for the prime implicants, and use this to select a simplest or minimal expression from those obtained by the Quine-McCluskey method. Applying this to the function of Example 4.5 we obtain two sums of prime implicants (in the order chosen): $E + I + M + L + G + A + F$, and $E + I + N + M + L + G + A + F$, having associated pairs $\langle 26, 13 \rangle$ and $\langle 29, 13 \rangle$ respectively. And by inspection, the first of these expressions is the simpler one.*

---

* Readers who are interested in more details on reducing the indeterminacy are referred to McCluskey's original paper [see Reference 4].

1. Use Karnaugh maps to minimize each of the following expressions.
   (a) $x + xy' + y'$.
   (b) $xy + xy'z + y(x' + z) + y'z'$.
   (c) $wx + xy + yz + zw + w'x'yz' + w'x'y'z$.
   (d) $wxy'z' + wxy'z + wxyz + wx'yz + w'x'yz + w'x'yz' + w'x'y'z'$ $+ w'xyz' + w'xy'z' + w'xy'z$.

2. Use the Quine-McCluskey method to simplify each of the following expressions.
   (a) The expression in Exercise 1c.
   (b) The expression in Exercise 1d.
   (c) $vw(x + y + xz') + v'x'z(wy' + x'(z' + v'y))$
   (d) $v'w'x'y'z' + v'w'x'yz' + v'wx'y'z' + v'wx'yz' + v'wxyz + vw'x'y'z'$ $+ vw'x'y'z + vw'xy'z + vw'xyz + vwx'y'z' + vwx'y'z + vwx'yz' +$ $vwxy'z + vwxyz$.

## 5. APPLICATIONS TO SWITCHING NETWORKS

In later chapters we shall discuss the relationships between Boolean algebra, logic, and the binary number system. We wish now to consider the use of Boolean algebra in the design of switching circuits.

We may associate each variable with a "gate" or valve in a line through which electricity, water, information, or some other quantity may flow. Physically, this gate might be a valve in an hydraulic line, an electrical relay, a transistor, a diode, a dispatcher at a warehouse, or any other device or person capable of permitting or prohibiting the flow of some quantity. Referring to Figure 5.1a, if there is an input at point $a$ then there is an output at point $b$ if and only if gate $x$ is closed.

We may then represent various logical functions in terms of such gates. For example, with an uncomplemented variable we associate a gate which is normally open; with a complemented variable, we associate a gate which

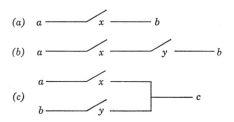

Figure 5.1. Basic switching circuitry.

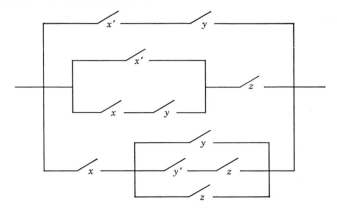

Figure 5.2. A switching circuit.

is normally closed, so that there will be an output if the variable is passive (gate left closed) and no output if the variable is active (gate opened). Similarly, in Figure 5.1b, if there is input at point $a$ then there will be output at point $b$ if and only if both of the gates $x$ and $y$ are closed. Thus this represents the term $xy$. And for there to be output at point $c$ in Figure 5.1c, given inputs at points $a$ and $b$, either gate $x$ or gate $y$ (or both) must be closed. Hence the expression $x + y$ is represented by this configuration.

*Example 5.1.* Let us consider the circuit shown in Figure 5.2 and determine what Boolean function it represents. The top line in this circuit represents the expression $x'y$ (remember that the $x'$ gate is normally closed while the $y$ gate is normally open). In the second line the far left portion is of the type shown in Figure 5.1c, and hence represents the expression $x' + xy$. Thus the whole line corresponds to the Boolean expression $(x' + xy)z$. Similarly, the bottom line of the figure corresponds to the expression $x(y + y'z + z)$. The whole circuit is again of the type shown in Figure 5.1c, and hence represents the Boolean expression

$$x'y + (x' + xy)z + x(y + y'z + z) \tag{5.1}$$

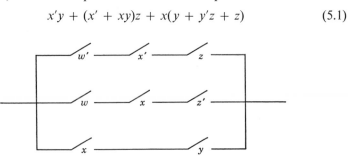

Figure 5.3. The switching circuit for $xy + wxz' + w'x'z$.

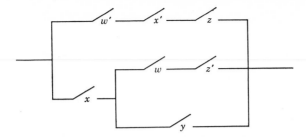

Figure 5.4. A circuit equivalent to Figure 5.3.

*Example 5.2.* The expression $w'x'z + wxz' + xy$ corresponds to the switching circuit in Figure 5.3. Notice that two of the gates are labeled $x$. We may in this case combine these two gates, obtaining the circuit in Figure 5.4, which involves one less gate. This corresponds to factoring the $x$ out of the last two products in the expression. Thus although the given expression is a minimal sum

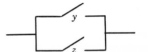

Figure 5.5. A circuit equivalent to Figure 5.2.

of products, if factoring is allowed a more simple expression can be found for the same function.

Because of the correspondence between Boolean functions and simple switching circuits, it is possible to apply the minimization techniques of Boolean algebra—graphical, tabular, and algebraic—to the problem of

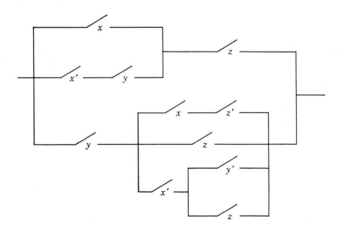

Figure 5.6.

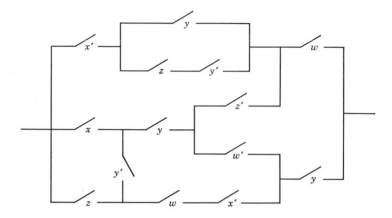

Figure 5.7.

simplifying a given switching circuit. For example, algebraic manipulation of (5.1) quickly reduces it to the expression $y + z$, and hence the circuit given in Figure 5.5 is equivalent to that in Figure 5.2. The reader may wish to check this equivalence directly.

One must be careful with the type of reduction used in Example 5.2, as it is possible to combine gates in ways which will introduce new "sneak"

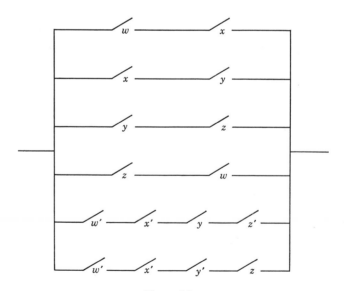

Figure 5.8.

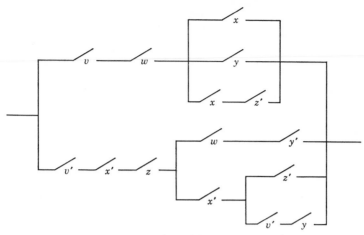

Figure 5.9.

paths into a circuit. Hence it is important to check the final circuit in order to be certain that it corresponds to the desired function.

EXERCISES

1. Draw the circuit diagram corresponding to each of the following expressions.
   (a) $xy + xy'z + y(x' + z) + y'z'$
   (b) $x + y + z'$
2. Write out the Boolean expression corresponding to each of the following circuit diagrams.
   (a) Figure 5.6.
   (b) Figure 5.7.
3. (a) For the circuit in Figure 5.8, find an equivalent circuit using 6 switches.
   (b) For the circuit in Figure 5.9, find an equivalent circuit using 9 switches.

## 6. FROM SETS TO LOGIC

Our particular concern with Boolean algebra lies in the fact that the subject provides a link between set theory and logic. At one end we have a two-element set algebra based, say, on the set $A = \{a\}$, with its two subsets $\varnothing$ and $A$. The three operation tables for this are given in Table 6.1.

**Table 6.1**

Operations in a two-element set algebra

| $\alpha \vartriangle \beta$ | | | | $\alpha \cup \beta$ | | | | $\alpha \cap \beta$ | | |
|---|---|---|---|---|---|---|---|---|---|---|
| $\alpha\backslash\beta$ | $\varnothing$ | $A$ | | $\alpha\backslash\beta$ | $\varnothing$ | $A$ | | $\alpha\backslash\beta$ | $\varnothing$ | $A$ |
| $\varnothing$ | $\varnothing$ | $A$ | | $\varnothing$ | $\varnothing$ | $A$ | | $\varnothing$ | $\varnothing$ | $\varnothing$ |
| $A$ | $A$ | $\varnothing$ | | $A$ | $A$ | $A$ | | $A$ | $\varnothing$ | $A$ |

Interpreting this as a Boolean algebra, we find that the two elements $\varnothing$ and $A$ must correspond to the 0 and 1 which are present in every Boolean algebra; a quick check verifies that $\varnothing$ corresponds to 0, and $A$ to 1. The operations for an abstract Boolean algebra on two elements are given in Table 6.2.

**Table 6.2**

Operations in a two-element Boolean algebra

| $\alpha \# \beta$ | | | | $\alpha + \beta$ | | | | $\alpha \cdot \beta$ | | |
|---|---|---|---|---|---|---|---|---|---|---|
| $\alpha\backslash\beta$ | 0 | 1 | | $\alpha\backslash\beta$ | 0 | 1 | | $\alpha\backslash\beta$ | 0 | 1 |
| 0 | 0 | 1 | | 0 | 0 | 1 | | 0 | 0 | 0 |
| 1 | 1 | 0 | | 1 | 1 | 1 | | 1 | 0 | 1 |

The logic most often studied is a two-valued logic, meaning that for purposes of the logic a statement may be either true (T) or false (F), but nothing else. The operation symbols used, $\not\equiv$, $\vee$, and $\wedge$, correspond to the Boolean operations $\#$, $+$, and $\cdot$ respectively, and the operation tables are given in Table 6.3.

**Table 6.3**

Operations in the propositional calculus

| $\alpha \not\equiv \beta$ | | | | $\alpha \vee \beta$ | | | | $\alpha \wedge \beta$ | | |
|---|---|---|---|---|---|---|---|---|---|---|
| $\alpha\backslash\beta$ | F | T | | $\alpha\backslash\beta$ | F | T | | $\alpha\backslash\beta$ | F | T |
| F | F | T | | F | F | T | | F | F | F |
| T | T | F | | T | T | T | | T | F | T |

# References

1. Arnold, B. H. *Logic and Boolean Algebra*. Prentice-Hall, Englewood Cliffs, New Jersey, 1962.
2. Boole, G. *An Investigation of the Laws of Thought*. Dover Publications, New York, 1958.
3. Flegg, H. G. *Boolean Algebra and Its Applications*. John Wiley and Sons, New York, 1964.
4. McCluskey, E. J., Jr. "Minimization of Boolean functions." *Bell System Technical Journal* **35**, 6 (1956), 1417–1444.
5. Miller, R. E. *Switching Theory. Vol. I: Combinational Circuits*. John Wiley and Sons, New York, 1965.
6. Quine, W. V. "The problem of simplifying truth functions." *American Mathematical Monthly*, **59** (1952), 521–531.
7. Quine, W. V. "A way to simplify truth functions," *American Mathematical Monthly*, **62** (1955), 627–631.
8. Rosenbloom, P. C. *The Elements of Mathematical Logic*. Dover Publications, New York, 1950.
9. Stoll, R. R. *Set Theory and Logic*. W. H. Freeman and Company, San Francisco, 1963.
10. Whitesitt, J. E. *Boolean Algebra and Its Applications*. Addison-Wesley Publishing Company, Reading, Mass., 1961.

# 3. The Propositional Calculus

## 1. BASIC NOTATION AND CONCEPTS

The propositional calculus is, literally, a method for calculating with propositions or sentences. We are concerned here with declarative sentences, which are either "true" or "false," and with methods of combining these and deducing other sentences from them. We will deal with these sentences in the abstract. That is, if we let "$p$" denote a sentence, we shall not ask what particular sentence $p$ denotes, but rather what happens if $p$, in a particular instance, denotes a true sentence, or a false one.

The first problem we encounter is that of defining the logical connectives which we shall use. If $p$ and $q$ denote simple declarative sentences, say $p$ denotes "the grass is green" and $q$ denotes "it is raining," we shall want to have symbols for handling such sentences as "$p$ is not true," "whenever $p$ then $q$," "$p$ and $q$," and so forth. In particular, whenever $p$ and $q$ are assigned truth values (that is, "true" or "false") we would like to be able to also assign truth values to the more complex sentences in a consistent manner which agrees with ordinary usage.

Let us begin with "it is not the case that $p$," or "$p$ is not true," symbolized by $\sim p$. It seems clear from ordinary usage that if $p$ is true we want $\sim p$ to be false, and if $p$ is false then $\sim p$ should be true. Thus we define the *negation* of $p$, $\sim p$ by Table 1.1, where the "T" denotes "true" and the "F" denotes "false."

**Table 1.1**

Definition of negation

| $p$ | $\sim p$ |
|-----|-----|
| T | F |
| F | T |

Another sentence whose intuitive truth value is clear is "$p$ and $q$": it should be true if and only if both $p$ and $q$ are true. Hence we define the *conjunction* of $p$ and $q$, $p \wedge q$, by Table 1.2.

**Table 1.2**

Definition of conjunction

| $p$ | $q$ | $p \wedge q$ |
|-----|-----|--------------|
| T | T | T |
| T | F | F |
| F | T | F |
| F | F | F |

The sentence "$p$ or $q$" normally is ambiguous. We use the word "or" both in an *inclusive* sense—"either $p$ is true or $q$ is true or both are true," and in an *exclusive* sense—"either $p$ is true or $q$ is true, but not both." In developing a formal system we wish to avoid such ambiguities, so we must then define two logical operators which correspond to the two meanings of "or". For the "inclusive or" we use the *disjunction*, $p \vee q$, defined by Table 1.3, and for the "exclusive or" we use the *nonequivalence*, $p \not\equiv q$, defined by Table 1.4. The reason for the term "nonequivalence" lies in the fact that $p \not\equiv q$ has the truth value T when $p$ and $q$ have different truth values, and F when they have the same truth value.

**Table 1.3**

Definition of disjunction

| $p$ | $q$ | $p \vee q$ |
|-----|-----|------------|
| T | T | T |
| T | F | T |
| F | T | T |
| F | F | F |

**Table 1.4**

Definition of nonequivalence

| $p$ | $q$ | $p \not\equiv q$ |
|-----|-----|------------------|
| T | T | F |
| T | F | T |
| F | T | T |
| F | F | F |

The four operations which we have defined provide the primary link back to the Boolean algebras of the last chapter. Negation, conjunction, disjunction, and nonequivalence correspond respectively to the Boolean operations complementation, multiplication, and the two additions + and #. Tables 1.2, 1.3, and 1.4 are appropriate revisions of Table 6.3 from Chapter 2.

Notice that we are also talking about mappings here: If we think of $p$ and $q$ not as denoting sentences, but as variables taking the value T and F, then $\vee$, for example, is a mapping from $\mathcal{T} \times \mathcal{T}$ into $\mathcal{T}$, where $\mathcal{T} = \{T, F\}$.

There are four elements in $\mathcal{T} \times \mathcal{T}$, namely $\langle T, T \rangle$, $\langle T, F \rangle$, $\langle F, T \rangle$, and $\langle F, F \rangle$. Any mapping from $\mathcal{T} \times \mathcal{T}$ into $\mathcal{T}$ must assign a value—either T or F—to each of these elements. Since each of these four elements may take either of the two values independently, there are $2 \times 2 \times 2 \times 2 = 16$ possible mappings of $\mathcal{T} \times \mathcal{T}$ into $\mathcal{T}$. The mappings which we have given and called disjunction, conjunction, and nonequivalence are just three of these. Table 1.5 lists all possible mappings.

### Table 1.5

Mappings of $\mathcal{T} \times \mathcal{T}$ into $\mathcal{T}$

| $p$ | $q$ | 1 | 2 | 3 | 4 | 5 | 6 | 7 | 8 | 9 | 10 | 11 | 12 | 13 | 14 | 15 | 16 |
|---|---|---|---|---|---|---|---|---|---|---|---|---|---|---|---|---|---|
| T | T | T | T | T | T | T | T | T | T | F | F | F | F | F | F | F | F |
| T | F | F | T | T | T | T | F | F | F | F | T | T | T | T | F | F | F | F |
| F | T | T | T | T | F | F | T | T | F | F | T | T | F | F | T | T | F | F |
| F | F | F | T | F | T | F | T | F | T | F | T | F | T | F | T | F | T | F |

Examining this table, we readily find the representation for the operations or connectives which we have already defined: disjunction ( $\vee$ ) is in column 2, conjunction ( $\wedge$ ) in column 8, and nonequivalence ( $\neq$ ) in column 10. We also note that $p$ is represented in column 4, $\sim p$ in column 13, $q$ in column 6, and $\sim q$ in column 11, column 1 is "universally true," and column 16 is "universally false." What of the remaining seven columns? It would seem plausible that these represent ways of making statements, and certainly we have some statements, such as "if $p$ then $q$" which we have not yet considered.

Let us note first that column $n$ in Table 1.5 is the negation of column $17-n$, in the same sense that $\sim p$ is the negation of $p$. Thus, as column 2 represents the disjunction $p \vee q$, column 15 represents *nondisjunction* or *nor* (not *or*): "it is not the case that $p$ or $q$ or both." The standard symbolism for this is the *Peirce's arrow*: $p \downarrow q$. Similarly, column 9 represents the negation of the conjunction (column 8), and is known as *nonconjunction* or *nand* (*not and*). It is generally indicated by *Sheffer's stroke*: $p \mid q$. And the negation of column 10 (nonequivalence) is the *biconditional* or *equivalence*, whose truth values are shown in column 7, and which is often symbolized by $p \equiv q$. It can be interpreted as "*p* if and only if *q*."

We are thus left with only four columns in Table 1.5 which do not yet represent statements, and two of these (12 and 14) are negations of the other two (5 and 3 respectively). The primary group of statements with which we have yet to deal are those of the form "if *p* then *q*" and "*p* whenever *q*." One source of difficulty here lies in the fact that with such statements we usually assume a causal relation. Thus many people would assert that the statement "if $2 + 2 = 4$, then the sun is shining" is non-sensical, since the value of $2 + 2$ has no apparent influence on the present behavior of the sun and the earth's atmosphere. However, we have in the logic presented here no way of indicating this causal connection. We can do no more than insist that our truth value assignment behave properly whenever such a causal relation exists.

Let us consider the situation of a candle in a sealed jar. Here we have a causal relation, namely a lack of oxygen in the jar causes the candle to fail to burn. This may be expressed in the form "if *p* then *q*": if (there is no oxygen in the jar) then (the candle does not burn). From physical experiments we would want to then regard the statement "if *p* then *q*" as true when both *p* and *q* have the value T, and false when *p* has the value T but *q* has the value F. Of the columns remaining, column 5 alone meets these requirements. Thus we use this column to symbolize the statement of the type "if *p* then *q*." The operation is known as the *conditional*, or (*material*) *implication*, and is written $p \supset q$. The *p* is called the *antecedent* and *q* the *consequent* of the conditional. The operation in column 3 is the same one with the roles of *p* and *q* interchanged, and thus provides the formal counterpart of the statement "*p* whenever *q*." It is often called *converse implication* and written $p \subset q$. The reader should also refer to the discussion in Section 4 of the present chapter to better understand the consequences of these definitions in argumentation.

The above discussion of the conditional is based entirely on what happens when *p* has the value T. The assignment of the value T to $p \supset q$ whenever *p* has the value F has bothered many people. Our development forced this: we had only one possible choice for the column representing "if *p* then *q*." Yet this is a poor justification for the remaining truth values for the conditional. The proper justification here lies in the fact that we are really more concerned with the form of an argument or statement than the truth of the conclusion. For example, the fact that *B* is a subset of *A* is characterized by the statement "if $b \in B$ then $b \in A$": a conditional. Since we wish to consider the empty set as a subset of any set, we want this characteristic statement to be true when $B = \varnothing$. But in this case, $b \in \varnothing$ is false for any $b$ in the universe, whereas $b \in A$ may or may not be true. Similar comments hold for logical arguments. Regardless of the truth value of the antecedent, if we are confident that the logical form of an argument is correct, then we can be assured that the conclusion is justified by the facts.

Table 1.6 shows the operation assignment which we have made. Not all of the operations which we have discussed are equally important. Sheffer's stroke and Peirce's arrow are important since any of the operations can be developed from either of these alone, and for this reason we shall discuss them in Section 8. However, these connectives represent rather uncommon modes of expression, and hence most of the work which we do will be concerned with the conjunction, disjunction, negation, conditional, and biconditional. As we shall see, it is not really necessary to use all of these; but they represent expressions so common in ordinary modes of discourse that it is convenient to have them all handy.

**Table 1.6**

The connectives of the propositional calculus

| "True" | $p \vee q$ | $p \subset q$ | $p$ | $p \supset q$ | $q$ | $p \equiv q$ | $p \wedge q$ |
|---|---|---|---|---|---|---|---|
| T | T | T | T | T | T | T | T |
| T | T | T | T | F | F | F | F |
| T | T | F | F | T | T | F | F |
| T | F | T | F | T | F | T | F |

| $p \mid q$ | $p \not\equiv q$ | $\sim q$ | $p \not\supset q$ | $\sim p$ | $p \not\subset q$ | $p \downarrow q$ | "False" |
|---|---|---|---|---|---|---|---|
| F | F | F | F | F | F | F | F |
| T | T | T | T | F | F | F | F |
| T | T | F | F | T | T | F | F |
| T | F | T | F | T | F | T | F |

EXERCISE

1. Write an expression in the symbolism of the propositional calculus corresponding to each of the following sentences.
   (a) The cat is blue and it eats green cheese.
   (b) Either tomorrow is Tuesday or it is Wednesday.

(c) Either the Yankees will win the pennant or they won't.

(d) If it is raining, I will stay home.

(e) I stay home whenever it is raining.

(f) If tomorrow is Tuesday and it does not rain, then either it will rain on Wednesday or the garden will dry up.

(g) The book should be about chemistry or biology, but if it is about biology it should be either about fungi or bacteria.

## 2. WELL-FORMED FORMULAS

Two things are apparent from Table 1.6. The first is that if these sixteen columns represent all possible assignments of truth values depending only on two arguments $p$ and $q$, then no matter how complex a sentence we construct from just two basic sentences, it is in some sense equivalent to one of these sixteen.

The other comment which is apparent is that we could construct similar tables for three variables $p$, $q$, and $r$, or even for more variables. However, such tables increase rapidly in size, with 256 ($= 2^{2^3}$) columns for three variables and 65,356 ($= 2^{2^4}$) columns for four variables. Fortunately, there is no need to construct such tables, because all such functions can be expressed in terms of the functions which we have already constructed, as we shall see in Section 6.

But how are we to construct these functions of several variables? What, for example, would $p \wedge q \vee r$ mean? Clearly, this particular question has two answers: either "$p$ and $q$, or $r$" or "$p$, and either $q$ or $r$." Thus we need some punctuation system to distinguish between these two meanings, together with rules for using the system. The resulting strings of symbols are called *well-formed formulas* (*wffs*) and are, intuitively, those symbol strings which make sense grammatically. The basic punctuation system which we shall use consists of parentheses.

Before developing the rules for well-formed formulas, we should pause to note that what we are developing is a highly symbolic and specialized language. In the course of our discussion we will want to make statements within this language, and statements about the language. We will want to prove theorems within the language, and theorems about the language. Thus we shall be talking on two distinct levels of discourse, called the "language" or "theory," and the "metalanguage" or "metatheory." We shall use the metalanguage to define precisely the modes of expression within the language; but we have (in the present work) no comparable meta-metalanguage in which to discuss the metalanguage. Hence our notions of sentence and proof within the metalanguage will be less precisely defined and largely intuitive.

We have already introduced certain symbols within our language or theory, namely the symbols for the various logical connectives. In addition we shall need as symbols *in the theory* the parentheses ( and ), and *propositional variables* $p$, $q$, $r$, $p_1$, $q_1$, $r_1$, $p_2$, .... Within the metalanguage we shall use the letters A, B, C, $A_1$, ... to denote well-formed formulas of the theory. Thus "(A) $\supset$ (B)" is not an expression in the theory, but rather an expression within the metatheory which denotes any desired member of the class of formulas in the theory obtained by substituting expressions in the theory for the letters A and B.

**Definition 2.1.** A *well-formed formula* (*wff*) of the propositional calculus is defined by the following rules:

1. A propositional variable standing alone is a wff.
2. If A is a wff, then $\sim$(A) is a wff.
3. If A and B are wffs, then the following are wffs:
   (A) $\wedge$ (B),   (A) $\vee$ (B),   (A) $\supset$ (B),   (A) $\equiv$ (B),   (A) $\subset$ (B),
   (A) $\not\supset$ (B),   (A) $\not\subset$ (B),   (A) $\not\equiv$ (B),   (A) $\mid$ (B),   and   (A) $\downarrow$ (B).
4. A symbol string is a wff if and only if its being so follows from finitely many applications of rules 1, 2, and 3.

*Example 2.1.* $((p) \supset ((q) \supset (r))) \supset (((p) \supset (q)) \supset ((p) \supset (r)))$ is a wff, as this analysis shows. Call this string $A_0$. Then

$$A_0 = (A_1) \supset (B_1), \quad \text{where}$$
$$A_1 = (p) \supset ((q) \supset (r))$$
$$B_1 = ((p) \supset (q)) \supset ((p) \supset (r))$$

$$A_1 = (A_2) \supset (B_2), \quad \text{where}$$
$$A_2 = p \quad \text{(a wff)}$$
$$B_2 = (q) \supset (r)$$

$$B_2 = (A_3) \supset (B_3), \quad \text{where}$$
$$A_3 = q \quad \text{(a wff)}$$
$$B_3 = r \quad \text{(a wff)}$$

$$B_1 = (A_4) \supset (B_4), \quad \text{where}$$
$$A_4 = (p) \supset (q)$$
$$B_4 = (p) \supset (r)$$

Then $A_4$ and $B_4$ are wffs by the same analysis as was used for $B_2$.

*Example 2.2.* $((p) \vee q)$ is not a wff, as this analysis shows. Call this string $C_0$. Then $C_0$ is not a propositional variable standing alone, nor is $C_0 = \sim(C_1)$ for some expression $C_1$. Hence if $C_0$ is a wff it must be so by rule 3. That is, $C_0 = (C_1)$ op $(C_2)$, where "op" is one of connectives listed in rule 3. In particular, the combination ") op (" must occur for $C_0$ to be a wff. But it does not.

The kind of analysis illustrated in these examples can be formalized into an algorithm or method for determining whether or not a given formula is well-formed. Basically such an algorithm must pair up the parentheses in the string properly and determine whether or not the rules apply. With the conventional notation used here, this process involves considerable memory work: in the first example given, if we scan from right to left, we must keep track of three right parentheses before we encounter a left parenthesis and can begin pairing. We shall investigate later a notation which avoids parentheses and other punctuation devices completely so as to simplify the problem.

The immediate impression upon looking at Example 2.1 is "too many parentheses." And indeed, for human consumption this is the case. Thus several conventions have been introduced to aid the person working in this area. First of all, the parentheses in $(p)$ appear redundant. Let us agree then that if A is a propositional variable standing alone, instead of "(A)" we will merely write "A". This simplifies the above example to

$$(p \supset (q \supset r)) \supset ((p \supset q) \supset (p \supset r))$$

A second convention which is used is that of a *hierarchy* or *order of precedence*. In arithmetic, for example, the expression $3 \times 4 + 5$ is ambiguous. Yet most people interpret this as $(3 \times 4) + 5$ rather than as $3 \times (4 + 5)$: they have been taught that multiplication takes precedence over addition. The order of precedence which we shall use is $\sim$, $\wedge$, $\vee$, $\supset$, $\equiv$. (The other connectives could be included, but they are rarely used.) Thus $p \wedge q \vee r$ means $(p \wedge q) \vee r$ rather than $p \wedge (q \vee r)$. The latter expression must include the parentheses. Similarly, $\sim p \supset q$ means $(\sim p) \supset q$, rather than $\sim (p \supset q)$. Furthermore, we shall agree that for two occurrences of the same connective, the leftmost takes precedence over the rightmost. Thus $p \wedge q \wedge r$ is $(p \wedge q) \wedge r$ rather than $p \wedge (q \wedge r)$; and $p \supset q \supset r \supset s$ is $((p \supset q) \supset r) \supset s$ rather than a formula with a different arrangement of parentheses. With this convention, our example could be written

$$p \supset (q \supset r) \supset (p \supset q \supset (p \supset r))$$

This is perhaps going too far: we are being frugal to the point where our formula is almost as difficult to read as the original version. Our prime criterion here is readability.

A third convention which is used is that of a system of dots to replace parentheses. Thus instead of (A) $\wedge$ (B) one may write A $\wedge$ .B, with the single dot replacing both pairs of parentheses. For the full use of such a system, one must introduce a hierarchy among the dot symbols, and use symbols which contain more than one dot. Full use of this system would cause our example to appear as

$$p \supset .q \supset r \supset :p \supset q \supset .p \supset r$$

We generally will not make extensive use of this system, but may use a single dot to indicate the main connective, thus writing

$$p \supset (q \supset r) \supset .(p \supset q) \supset (p \supset r)$$

as the "most readable" version of our formula.* Notice that what we now have is *not* a well-formed formula as we have defined them, but rather an abbreviation for one. In using any convention such as those which we have described, we must, of course, be able to reconstruct the fully parenthesized expression from its abbreviation.

<h2 style="text-align:center">EXERCISES</h2>

1. Which of the following are well-formed formulas?
    (a) $((p) \wedge (\sim(q))) \supset ((\sim(\sim(p))) \equiv (q))$
    (b) $((p) \wedge (q)) \supset (\sim(q) \wedge (p) \vee (r))$
    (c) $(p) \supset (((q) \sim (r)) \equiv ((p) \vee (\sim((q) \wedge (r)))))$
    (d) $((((s) \downarrow ((p) \mid (q))) \equiv (r)) \supset (\sim((p) \mid (q)))) \wedge ((r) \supset (((\sim(s))$
        $\equiv (\sim((p) \vee (r)))) \supset (((s) \equiv (\sim(p))) \wedge (q))))$
    (e) $(((p) \supset (\sim(q))) \equiv ((r) \supset ((\sim(q)) \vee (q))))$
    (f) $(p) \mid ((q) \mid ((r) \mid ((p) \mid ((r) \mid ((p) \mid (q))))))$
2. Using the conventions which we have established, expand each of the following into well-formed formulas.
    (a) $p \equiv q \equiv r$
    (b) $\sim p \vee r \equiv p \supset q \wedge \sim q \supset r$
    (c) $\sim p \vee r \equiv (p \supset q) \wedge (\sim q \supset r)$
    (d) $\sim(p \wedge q \vee r) \supset p \equiv q$
    (e) $p \vee r \supset p \vee q \supset q \vee r$

## 3. TRUTH TABLES

We now consider the problem of evaluating a logical expression, taking as our ideogenic formula

$$p \wedge q \supset .p \vee \sim r \tag{3.1}$$

* For further information on the dot hierarchy, see Rosenbloom [5], p. 32.

(that is, $(p \wedge q) \supset (p \vee \sim r)$). The truth value of this expression clearly depends on those of $p$, $q$, and $r$, and the expression represents a *truth function* in the sense which we previously indicated: namely, this expression denotes a function from $\mathcal{T} \times \mathcal{T} \times \mathcal{T}$ into $\mathcal{T}$. Since $\mathcal{T} \times \mathcal{T} \times \mathcal{T}$ contains only eight ($2 \times 2 \times 2$) points, we may give a complete list of the values of this function in a *truth table*. We first give an extended truth table, then a compressed form, and finally what may be called a "sparse" table. In the extended table (Table 3.1) the columns to the right of the

**Table 3.1**

Evaluation of the expression (3.1)

| $p$ | $q$ | $r$ | $p \wedge q$ | $\sim r$ | $p \vee \sim r$ | $p \wedge q \supset \cdot p \vee \sim r$ |
|---|---|---|---|---|---|---|
| T | T | T | T | F | T | T |
| T | T | F | T | T | T | T |
| T | F | T | F | F | T | T |
| T | F | F | F | T | T | T |
| F | T | T | F | F | F | T |
| F | T | F | F | T | T | T |
| F | F | T | F | F | F | T |
| F | F | F | F | T | T | T |

vertical line give a dissection of the formula into smaller parts which can be evaluated directly. The evaluation in the table then proceeds from left to right, with the final column given the truth values for the function represented by the given expression.

In the compressed table (Table 3.2) the same evaluation is done, but the results are indicated directly below the connective involved. Thus considerable extra writing is avoided. The final column representing the values of the function is outlined for clarity.

In developing a "sparse" truth table we take advantage of our knowledge of the truth values of the various connectives. For example, the main connective in the given formula is a conditional, and we know that it has the value *true* whenever the antecedent is *false*. The antecedent, in our case, is a conjunction, and is false whenever either of the operands $(p, q)$ is false. Similarly, the conditional is true if the consequent is true. And the consequent, being a disjunction, is true if either of the operands $(p, \sim r)$ is true. Thus for this particular example, we can determine the entire truth value assignment from a knowledge of the value of $p$, as is shown in Table 3.3.

**Table 3.2**

Compressed truth table for expression (3.1)

| $p$ | $q$ | $r$ | $p \wedge q$ | $\supset$ | $.p \vee$ | $\sim r$ |
|-----|-----|-----|------|---|---|---|
| T | T | T | T | T | T | F |
| T | T | F | T | T | T | T |
| T | F | T | F | T | T | F |
| T | F | F | F | T | T | T |
| F | T | T | F | T | F | F |
| F | T | F | F | T | T | T |
| F | F | T | F | T | F | F |
| F | F | F | F | T | T | T |

A truth table, then, is a tabular array showing the truth value of a wff for each assignment of truth values to its component propositions. In a sparse table one often finds several lines of the table condensed into one, as was indicated in Table 3.3. Thus the single line "T – –" for $p, q, r$ represents the four lines T T T, T T F, T F T, and T F F.

**Definition 3.1.** If for each assignment of values to its variables a wff has the value T it is called a *tautology*; if it always has the value F it is called a *contradiction*.

Hence the final or principal column of the truth table of a tautology consists entirely of T's, and that of a contradiction consists entirely of F's. Clearly a wff A is a tautology if and only if its negation $\sim$A is a contradiction. If A is a tautology, we write ⊢A.

*Example 3.1.* Let us consider the wff (3.2) in four variables.

$$p \supset (\sim(q \vee \sim r) \wedge (p \vee s)) \equiv .(\sim s \supset p) \wedge (r \supset \sim q) \qquad (3.2)$$

The evaluation of this expression is given in Table 3.4, where the rows are listed in the order in which they were examined. In working with a sparse truth table, it is prudent to examine the formula quite carefully before

**Table 3.3**

Sparse truth table for expression (3.1)

| $p$ | $q$ | $r$ | $p \wedge q$ | $\supset .$ | $p \vee \sim r$ | |
|-----|-----|-----|------|-----|---|---|
| T | – | – | – | T | T | |
| F | – | – | F | T | – | |

beginning to construct the table. In the present case the main connective is a biconditional, and hence does not immediately suggest whether or not it

**Table 3.4**

Truth table for expression (3.2)

| p | q | r | s | p ⊃ | (~(q | ∨ | ~r) ∧ (p | ∨ | s)) | ≡ .(~s | ⊃ p) | ∧ (r | ⊃ ~q) |
|---|---|---|---|---|---|---|---|---|---|---|---|---|---|
| F | – | – | F | T | | | | | | F | F | F | |
| F | F | – | T | T | | | | | | T | T | T | T |
| F | – | F | T | T | | | | | | T | T | T | T |
| F | T | T | T | T | | | | | | F | | F | F |
| T | – | F | – | F | F | T | F | | | F | T | T | T |
| T | T | T | – | F | F | T | F | | | T | | F | F |
| T | F | T | – | T | T | F | T | T | | T | T | T | T |

would be easier to find true values than false. However, we note that the left side of the formula is a conditional whose antecedent is *p*. Thus whenever *p* has the value F the whole left side has the value T regardless of the values of the other variables. Turning to the right side with the condition that *p* have the value F, we note that giving *s* the value F causes the whole right side of the expression to be false, and hence forces the expression itself to have the value F. Thus we have the first line in the table. The other lines are generated similarly. Notice that a sixteen-line truth table is thus compressed to seven lines, with one line (F F F T) represented twice. The given expression is neither a tautology nor a contradiction since its truth value can be either T or F depending on the values of the variables.

EXERCISES

In each of the following exercises, write out the truth table and determine whether the formula is a tautology, a contradiction, or neither.

1. $p \supset (p \supset p) \supset .p$
2. $(p \wedge \sim q) \supset (\sim \sim p \equiv q)$
3. $p \equiv q \equiv r$
4. $(p \wedge q) \supset (\sim q \wedge (p \vee r))$
5. $\sim (p \wedge q \vee r) \supset p \equiv q$
6. $(p \supset \sim q) \equiv (r \supset (\sim q \vee p))$
7. $(p \equiv q) \equiv r \equiv .p \equiv (q \equiv r)$

8. $p \lor r \supset p \lor q \supset q \lor r$

9. $p \supset .(q \lor r) \equiv (p \lor (\sim q \land r))$

10. $\sim p \lor r \equiv p \supset q \land \sim q \supset r$

11. $\sim p \lor r \equiv (p \supset q) \land (\sim q \supset r)$

12. $p \mid (q \mid (r \mid (p \mid (r \mid (p \mid q)))))$

13. $p \supset (q \land \sim r) \equiv .(p \lor \sim q) \land (r \supset p)$

14. $\sim((p \supset \sim((q \lor \sim r) \lor (p \equiv s))) \lor \sim(\sim p \equiv ((q \supset r) \land \sim s)))$

15. $(s \downarrow (p \mid q) \equiv r) \supset \sim (p \mid q) \land .r \supset ((\sim s \equiv \sim (p \lor r)) \supset ((s \equiv \sim p) \land q))$


## 4. ARGUMENTATION AND EVALUATION

The basic problem in truth-functional calculus, as an applied art, is to represent and evaluate the statements and arguments of everyday life. Consider, for example, the following statement:

If the drug passes both test A and test B, then the company will market it if and only if it can be produced and sold profitably and the government does not intervene.

An appropriate logical analysis of this statement might begin with the symbolism

$$a \land b \supset (m \equiv p \land \sim g) \tag{4.1}$$

where the letters denote the following statements:

$a$: The drug passes test A.
$b$: The drug passes test B.
$m$: The company will market the drug.
$p$: The drug can be produced and sold profitably.
$g$: The government intervenes.

The basic question to be asked is then under what conditions this statement is true. We have five propositional variables occurring here, and since each of these may be assigned the value T or F independently, there are $2^5 = 32$ lines in the full truth table. Table 4.1 gives the truth values for this expression.

One should notice first that the first two lines of this sparse table represent the situation if the drug fails one or both of the tests *and that in this case the proposition is true.* This does not imply that the company might go ahead and market the drug anyway, but rather recognizes the fact that no claim was made about what would be done in this case: "*If the drug passes both test A and test B, then . . . .*"

Turning then to the remainder of the table we find that in half of the lines the statement is false—that is, even though the drug passed the two tests, for some reason (for example, the marketing of an unprofitable product) the conclusion of the statement does not hold. Those lines remaining for which the statement is true represent the situation which should hold if both tests are passed.

In extended arguments the same sort of analysis may be applied. Given a set of statements, we may examine each to determine for which combinations of truth values of its propositional variables it is true, and for which it is false. Generally we want also to examine the relation between various statements using certain rules of inference. This will be explored more fully in the next section.

**Table 4.1**

Truth table for expression (4.1)

| $a$ | $b$ | $m$ | $p$ | $g$ | $a \wedge b$ | $\supset$ | $(m$ | $\equiv$ | $p$ | $\wedge$ | $\sim g)$ |
|---|---|---|---|---|---|---|---|---|---|---|---|
| F | – | – | – | – | F | T | | | | | |
| – | F | – | – | – | F | T | | | | | |
| T | T | T | F | – | T | F | F | | F | | |
| T | T | T | – | T | T | F | F | | F | | |
| T | T | F | T | F | T | F | F | | T | | |
| – | – | T | T | F | | T | T | | T | | |
| – | – | F | F | – | | T | T | | F | | |
| – | – | F | – | T | | T | T | | F | | |

EXERCISES

Analyze each of the following statements to determine under what conditions it is either true or false.

1. The painting is of value only if it is at least 600 years old, or it is less than 600 years old and it was painted with hand-ground pigments, or, irrespective of age, it was painted on silk or on linen.
2. We shall go to the meeting if and only if there are funds available, and either the meeting does not interfere with our work or there is a session at the meeting which is directly related to our work.
3. It is possible for a person to drown in water three inches deep provided that either he is unconscious and lying face down in the water, or that his mustache is on fire and he is trying to extinguish the blaze.

## 5. LOGICAL EQUIVALENCE AND LOGICAL CONSEQUENCE

**Definition 5.1.** Two statements A and B are *logically equivalent* (A eq B) if and only if they have the same truth table.

This does not imply that A and B "mean" the same thing: "if $x$ is an integer, then $2x$ is an integer" and "if the moon is made of green cheese, then it is raining" both have the same *logical* form ($p \supset q$), and hence are logically equivalent. There is implicit in this definition the idea that the variables occurring in A are the same as those occurring in B, for if the value of A depends on the value of, say, $p$, but B does not contain an occurrence of $p$, then generally A and B will not be logically equivalent.

**Theorem 5.1.** A eq B if and only if $\vDash A \equiv B$.

*Proof.* Suppose first that $\vDash A \equiv B$, that is, that independently of the assignment of truth values to the variables in $A \equiv B$, this formula has the value T. Then for each particular assignment of truth values to these variables, either A has the value T and B has the value T, or both have the value F. Thus A and B have the same truth table, and hence A eq B.

Now suppose that A eq B, and consider any particular assignment of truth values to the variables in A and B. If, for this particular assignment, A has the value T, then so also B has the value T, and hence $A \equiv B$ has the value T. Similarly, if for this assignment A has the value F, then B also has the value F, and hence $A \equiv B$ has the value T. Thus, since $A \equiv B$ always has the value T, $\vDash A \equiv B$.

If we say that "B is a logical consequence of A" we intend to convey the idea that whenever A is true, then B must necessarily be true. This prompts the following definition and theorem.

**Definition 5.2.** B is a *logical consequence* of A (A $\vDash$ B) if for each assignment of truth values to the variables of A (and B) such that A has the value T, then B also has the value T.

**Theorem 5.2.** A $\vDash$ B if and only $\vDash A \supset B$.

*Proof.* The proof of this theorem is exactly analogous to the preceding proof, and is left to the reader.

The obvious generalization of this line of thought is to replace A by a series of formulas and say that $A_1, \ldots, A_n \vDash B$ if and only if whenever $A_1, \ldots, A_n$ are all true, then B is also true. The above theorem generalizes also.

**Theorem 5.3.** $A_1, \ldots, A_n \vDash B$ if and only if $\vDash A_1 \supset (\ldots \supset (A_n \supset B)\ldots)$.

**Corollary.** $A_1, \ldots, A_n \vDash B$ if and only if $\vDash A_1 \wedge \cdots \wedge A_n \supset B$.

*Proof.* It is easily established that

$$A_1 \supset (\cdots (A_n \supset B) \cdots) \text{ eq } A_1 \wedge \cdots \wedge A_n \supset B$$

The result then follows from Theorem 5.3.

**Corollary.** If  A, B ⊢C,  then  B, A ⊢C.  If  $A_1, A_2, \ldots, A_n$ ⊨B,  then $A_{i_1}, A_{i_2}, \ldots, A_{i_n}$ ⊨B,  where  $i_1, i_2, \ldots, i_n$  is  any  permutation  of  the integers $1, 2, \ldots, n$.

Thus one may move back and forth between the ideas of logical conse-quence and tautology to a certain extent. In particular, one may establish tautologies and use these to generate informal *rules of inference*. One can then use these rules of inference for further investigations, knowing that one can always generate the tautologies behind these. This method has the practical advantage that much of the truth table checking can be eliminated.

*Example 5.1.* We illustrate with the development of the rule of inference known as *modus ponens*. The first step is to establish ⊨A ∧ (A ⊃ B) ⊃ .B by Table 5.1.

**Table 5.1**

Truth table for A ∧ (A ⊃ B) ⊃ .B

| A | B | A ∧ | (A ⊃ B) | ⊃ .B |
|---|---|-----|---------|------|
| T | T | T | T | T |
| T | F | F | F | T |
| F | T | F | T | T |
| F | F | F | T | T |

Then using the preceding corollary, we may assert that A, A ⊃ B ⊨B. That is, on the assumption of the formulas A and A ⊃ B, we may establish B. In particular, *if ⊨A and ⊨A ⊃ B, then ⊨B.* (Notice that this is the first line of our truth table.)

*Example 5.2.* To illustrate the analysis of an argument, we examine the following problem.*

A real revolt involves resolution, responsibility, and reformation. An unsuccessful retort means that recognition means that a lack of relativity involves a lack of reformation. Resolution involves an unsuccessful retort. Responsibility means that recourse involves recognition. Relativity implies the absence of recourse. There is recourse. Therefore there is not any real revolt. Is this conclusion valid?

* From *More Problematical Recreations*, Litton Industries. Used with permission of Litton Industries.

We use the following symbolism:

|                    |     |
|--------------------|-----|
| real revolt        | $V$ |
| resolution         | $S$ |
| responsibility     | $P$ |
| reformation        | $M$ |
| unsuccessful retort | $R$ |
| recognition        | $G$ |
| relativity         | $L$ |
| recourse           | $C$ |

The argument then becomes the following:

$A_1$:   $V \supset S \wedge P \wedge M$
$A_2$:   $R \supset (G \supset (\sim L \supset \sim M))$
$A_3$:   $S \supset R$
$A_4$:   $P \supset (C \supset G)$
$A_5$:   $L \supset \sim C$
$A_6$:   $C$
_____
B:     $\sim V$          [conclusion (?)]

This is purported to be an argument of the form $A_1, \ldots, A_6 \vdash B$. Thus we must examine the conditions under which $A_1, \ldots, A_6$ are true. We shall use the notation $A \Rightarrow V$, where $V$ is either T or F, to say that A is given the truth value $V$.

1. $A_6 \Rightarrow$ T if and only if $C \Rightarrow$ T.
2. Because of 1, $A_5 \Rightarrow$ T if and only if $L \Rightarrow$ F.
3. Now suppose that $V \Rightarrow$ T. Then $A_1 \Rightarrow$ T if and only if $S \Rightarrow$ T, $P \Rightarrow$ T, and $M \Rightarrow$ T.
4. From 1 and 3, $A_4 \Rightarrow$ T if and only if $G \Rightarrow$ T.
5. From 3, $A_3 \Rightarrow$ T if and only if $R \Rightarrow$ T.
6. Then by 2, 4, and 5, $A_2 \Rightarrow$ T if and only if $M \Rightarrow$ F.
7. But this contradicts the assignment given to $M$ in 3. Hence our supposition must be wrong, and thus $V \Rightarrow$ F.
8. Thus the conclusion is valid.

### EXERCISES

1. For each of the following pairs, show that A eq B.

|       | A | B |
|-------|---|---|
| (a) | $\sim\sim p$ | $p$ |
| (b) | $p \equiv q \wedge \sim q$ | $\sim p$ |
| (c) | $p \wedge q \equiv p$ | $p \supset q$ |
| (d) | $(p \wedge (q \supset r)) \vee (\sim p \wedge q)$ | $(\sim p \vee \sim q \vee r) \wedge (p \vee q)$ |
| (e) | $(\sim p \wedge (r \supset q)) \vee (p \wedge \sim(q \supset r))$ | $p \wedge q \wedge \sim r \vee \sim p \wedge q \vee \sim p \wedge \sim r$ |

2. For each of the following pairs, show that A ⊨B.

| | A | B |
|---|---|---|
| (a) | $p$ | $p \supset q \supset q$ |
| (b) | $p \supset q$ | $(q \supset r) \supset (p \supset r)$ |
| (c) | $p \supset q$ | $(\sim p \supset q) \supset q$ |
| (d) | $p \lor (q \lor r)$ | $(q \lor (p \lor r)) \lor p$ |
| (e) | $(((p \supset q) \supset (\sim r \supset \sim s)) \supset r) \supset t$ | $(t \supset p) \supset (s \supset p)$ |

3. Show that $A_1$, $A_2$, $A_3$, $A_4$ ⊨B, where

$A_1$ is $p \supset q \lor r$

$A_2$ is $q \land \sim s \supset p \lor t$

$A_3$ is $\sim q \equiv \sim(p \lor (s \land t))$

$A_4$ is $(s \supset \sim t) \lor (p \land \sim q)$

and   B is $(p \equiv q) \land \sim(r \land s \land t)$

4. Analyze the following situation.*

Wishes are horses, provided that horses cannot fly. Beggars will not ride, provided that wishes are not horses. If it cannot be the case that both beggars will ride and wishes are nonequine, then horses can fly. If the inability of horses to fly, and the nonriding of beggars, cannot be set up as valid alternatives, then beggars are not always rich. But beggars *will* ride. Are beggars rich?

## 6. NORMAL FORMS

Earlier in this chapter we made the claim that all logical functions of more than two variables could be represented by using only the binary functions. In this section we develop certain standard forms for functions, and in the course of this development show that this earlier claim is indeed justified.

We have a wealth of connectives with which to work—ten binary connectives plus negation. Thus it is easily conceivable that if A and B are complex formulas involving many variables and a variety of connectives, the task of establishing whether or not A is logically equivalent to B could be tedious, and difficult to implement on a computer. We would like to develop a standard form to simplify this task. Ideally, it should be possible to easily transform A and B into standard forms A' and B', such that A eq B could be established or disproved by a quick scan of A' and B'. This is possible, although "easily" might not be considered a proper word.

* From *More Problematical Recreations*, Litton Industries. Used with permission of Litton Industries.

There is another reason also for a standard form, namely, given a wff A, to select a wff B, B eq A, which is in some sense minimal. For example, the formula $(p \wedge q) \supset (\sim p \vee (p|\sim q)) \neq .\sim p \supset ((q \subset p) \downarrow \sim q)$ must simplify to one of the sixteen forms heading Table 1.6 since it involves only two variables. (In fact, it is equivalent to $p \downarrow q$.) Similarly, a formula in $n$ variables will have a "simplest" form, and the standard forms provide a systematic way to find the minimal forms.

The forms which we shall discuss in this section are related to the normal forms introduced in Section 3 of Chapter 2. In fact, since the logic which we have been developing constitutes a set of Boolean functions, all the techniques discussed in the previous chapter may be used in the propositional calculus. Conversely, the discussion here may be applied *mutatis mutandis* to the Boolean functions previously discussed, the Boolean sum and product corresponding respectively to disjunction and conjunction.

Of the several *standard* or *canonical* or *normal forms* which have been developed, we shall consider only four closely related forms which are easily developed from either a truth table or a formula. These represent a formula in a form

$$(p_1 \wedge q_1 \wedge \cdots) \vee (p_2 \wedge q_2 \wedge \cdots) \vee \cdots$$

or

$$(p_1 \vee q_1 \vee \cdots) \wedge (p_2 \vee q_2 \vee \cdots) \wedge \cdots$$

We suppose that the propositional variables are given in some order, which we shall call the *alphabetical order*, say $p, q, r, p_1, q_1, r_1, \ldots$.

**Definition 6.1.** A formula is an *elementary conjunct* if it is of the form $A_1 \wedge A_2 \wedge \cdots \wedge A_n$, where

1. each $A_i$ is either a propositional variable or the negation of a propositional variable;
2. no propositional variable occurs in more than one $A_i$;
3. those propositional variables which occur are in alphabetic order— that is, if $i < j$ then the propositional variable in $A_i$ precedes that in $A_j$ in the alphabet.

A formula is in *disjunctive normal form* (DNF) if it is of the form $B_1 \vee B_2 \vee \cdots \vee B_m$, where

1. each $B_i$ is an elementary conjunct;
2. no two of the $B_i$ are the same;
3. If $B_i = A_{i1} \wedge A_{i2} \wedge \cdots \wedge A_{in_i}$  and
   $B_j = A_{j1} \wedge A_{j2} \wedge \cdots \wedge A_{jn_j}$  and
   $A_{i1} = A_{j1}, A_{i2} = A_{j2}, \ldots, A_{i,k-1} = A_{j,k-1}$
and  $A_{ik} \neq A_{jk}$,  then
$i < j$ if and only if one of the following holds:

(a) $A_{ik}$ does not exist (that is, $k - 1 = n_i$).
(b) $A_{ik} = p$ and $A_{jk} = \sim p$ for some propositional variable $p$.
(c) the propositional variable occurring in $A_{ik}$ precedes that occurring in $A_{jk}$.

A wff is in *full disjunctive normal form* (FDNF) if it is in DNF and each propositional variable which occurs in any one of the elementary conjuncts occurs in all of the elementary conjuncts.

The terms *elementary disjunct, conjunctive normal form* (CNF), and *full conjunctive normal form* (FCNF) are similarly defined with the disjunction and conjunction interchanged.

Notice that this definition completely specifies the normal forms, down to the order in which the various terms occur. Thus once two formulas have been written in one of the full normal forms a single direct comparison of the two forms will establish whether or not the formulas represent the same function. Notice also that the simplest expression for a tautology, $p \lor \sim p$, is a DNF but not a CNF since $p$ and $\sim p$ cannot both occur in the same elementary disjunct. In general, a tautology does not have a CNF, and a contradiction has no DNF.

There are two ways to determine a normal form corresponding to a given formula. One is to effect a series of transformations based on logical equivalences such as are given in Table 6.1.

**Table 6.1**

Basic logical equivalences

| | | |
|---:|:---:|:---|
| $p \land p$ | eq | $p$ |
| $p \lor p$ | eq | $p$ |
| $p \land (q \lor r)$ | eq | $(p \land q) \lor (p \land r)$ |
| $p \lor (q \land r)$ | eq | $(p \lor q) \land (p \lor r)$ |
| $\sim \sim p$ | eq | $p$ |
| $\sim (p \lor q)$ | eq | $\sim p \land \sim q$ |
| $\sim (p \land q)$ | eq | $\sim p \lor \sim q$ |
| $p \land (p \lor q)$ | eq | $p$ |
| $p \lor (p \land q)$ | eq | $p$ |
| $p \supset q$ | eq | $\sim p \lor q$ |
| $p \equiv q$ | eq | $(p \land q) \lor (\sim p \land \sim q)$ |
| $p \subset q$ | eq | $p \lor \sim q$ |
| $p \mid q$ | eq | $\sim p \lor \sim q$ |
| $p \downarrow q$ | eq | $\sim p \land \sim q$ |
| $p \not\supset q$ | eq | $p \land \sim q$ |
| $p \not\equiv q$ | eq | $(\sim p \lor \sim q) \land (p \lor q)$ |
| $p \not\subset q$ | eq | $\sim p \land q$ |

*Example 6.1*   $\sim (p \wedge q) \supset \sim p \vee (p \mid \sim q)$
$\sim\sim(p \wedge q) \vee (\sim p \vee (\sim p \vee \sim\sim q))$
$(p \wedge q) \vee (\sim p \vee \sim p \vee q)$
$(p \wedge q) \vee \sim p \vee q$    (DNF, 3 elementary conjuncts)
$\sim p \vee q$    (DNF, 2 elementary conjuncts, or CNF, 1 elementary disjunct.)

*Example 6.2*   $(\sim p \supset r) \wedge (q \equiv p)$
$(\sim\sim p \vee r) \wedge ((q \wedge p) \vee (\sim q \wedge \sim p))$
$(p \vee r) \wedge ((p \wedge q) \vee (\sim p \wedge \sim q))$
$((p \vee r) \wedge p \wedge q) \vee ((p \vee r) \wedge \sim p \wedge \sim q)$
$(p \wedge q) \vee (((p \wedge \sim p) \vee (r \wedge \sim p)) \wedge \sim q)$
$(p \wedge q) \vee (\sim p \wedge \sim q \wedge r)$    (DNF)
$(p \vee (\sim p \wedge \sim q \wedge r)) \wedge (q \vee (\sim p \wedge \sim q \wedge r))$
$(p \vee \sim p) \wedge (p \vee \sim q) \wedge (p \vee r) \wedge (q \vee \sim p)$
$\qquad\qquad\qquad\qquad \wedge (q \vee \sim q) \wedge (q \vee r)$
$(p \vee \sim q) \wedge (p \vee r) \wedge (\sim p \vee q) \wedge (q \vee r)$    (CNF)

The second method for obtaining normal forms for a formula is by examination of the truth table. In this method we obtain initially the FDNF, and then use this to generate the other normal forms. This method is particularly useful if we do not know the formula, but only its truth table. Suppose that A is a wff and we know that for a particular assignment of truth values to its variables, A has the value true. Then that assignment must correspond to one *or* another of the lines of A's truth table for which A has the value T, and whichever line it is, *all* of the variables must have the assignments in that line. This leads to the FDNF.

*Example 6.3.* Let us consider the same formula used in the previous example, $(\sim p \supset r) \wedge (q \equiv p)$. Table 6.2 gives the truth values for this wff.

**Table 6.2**

Truth table for $(\sim p \supset r) \wedge (q \equiv p)$

| $p$ | $q$ | $r$ | $(\sim p \supset r)$ | $\wedge$ | $(q \equiv p)$ |
|---|---|---|---|---|---|
| T | T | T | T | T | T |
| T | T | F | T | T | T |
| T | F | T | T | F | F |
| T | F | F | T | F | F |
| F | T | T | T | F | F |
| F | T | F | F | F | F |
| F | F | T | T | T | T |
| F | F | F | F | F | T |

Since the value is T for the first, second, and seventh lines of this table the FDNF is the disjunction of elementary conjunctions representing those lines, $(p \wedge q \wedge r) \vee (p \wedge q \wedge \sim r) \vee (\sim p \wedge \sim q \wedge r)$. By use of an appropriate transformation this reduces to the DNF given in Example 6.2.

*Example 6.4.* We now consider an example in which the function is known only by its values. That is, we have no given formula to represent the function, but only the values as given in Table 6.3. In such a case we can quickly represent the function by a formula in FDNF. For the given function that formula is

$$(p \wedge \sim q \wedge r) \vee (\sim p \wedge q \wedge r) \vee (\sim p \wedge q \wedge \sim r) \vee (\sim p \wedge \sim q \wedge \sim r)$$

By use of appropriate transformations, different forms for the function can be derived, as for example

$$(\sim p \wedge (r \supset q)) \vee (p \wedge \sim q \wedge r)$$

or
$$(\sim q \wedge (p \equiv r)) \vee (\sim p \wedge q)$$

**Table 6.3**

The truth table for Example 6.4

| $p$ | $q$ | $r$ | $f(p, q, r)$ |
|-----|-----|-----|--------------|
| T | T | T | F |
| T | T | F | F |
| T | F | T | T |
| T | F | F | F |
| F | T | T | T |
| F | T | F | T |
| F | F | T | F |
| F | F | F | T |

Once we have the full disjunctive normal form of a formula, it is a simple matter to obtain the full conjunctive normal form (FCNF). Let A be a wff; then $\sim$A has the value T if and only if A has the value F. Hence the FDNF for $\sim$A is obtained from the truth table for A in the same manner as is the FDNF for A, except that the rows with value F are used. From this form we may obtain the FCNF for A by using DeMorgan's laws:

$$\sim(A \wedge B) \text{ eq } \sim A \vee \sim B, \qquad \sim(A \vee B) \text{ eq } \sim A \wedge \sim B,$$

and the law of double negation: $\sim\sim A$ eq A. If the FDNF for $\sim A$ is $B_1 \vee B_2 \vee \cdots \vee B_n$, then

$$A \text{ eq } \sim\sim A$$
$$\text{eq } \sim(B_1 \vee B_2 \vee \cdots \vee B_n)$$
$$\text{eq } \sim B_1 \wedge \sim B_2 \wedge \cdots \wedge \sim B_n.$$

But a typical $B_i$ is $C_{i1} \wedge C_{i2} \wedge \cdots \wedge C_{in_i}$, so that $\sim B_i$ eq $\sim C_{i1} \vee \sim C_{i2} \vee \cdots \vee \sim C_{in_i}$, and thus by judicious use of the law of double negation we have the desired FCNF for A.

*Example 6.5.* Let us consider again the formula $(\sim p \supset r) \wedge (q \equiv p)$, whose FDNF was found in Example 6.3. Call this formula A. Then the FDNF for $\sim A$ is

$$(p \wedge \sim q \wedge r) \vee (p \wedge \sim q \wedge \sim r) \vee (\sim p \wedge q \wedge r) \vee (\sim p \wedge q \wedge \sim r)$$
$$\vee (\sim p \wedge \sim q \wedge \sim r) \quad (6.1)$$

Thus

$$A \text{ eq } \sim(\sim A)$$
$$\text{eq } \sim(p \wedge \sim q \wedge r) \wedge \sim(p \wedge \sim q \wedge \sim r) \wedge \sim(\sim p \wedge q \wedge r)$$
$$\wedge \sim(\sim p \wedge q \wedge \sim r) \wedge \sim(\sim p \wedge \sim q \wedge \sim r)$$
$$\text{eq } (\sim p \vee q \vee \sim r) \wedge (\sim p \vee q \vee r) \wedge (p \vee \sim q \vee \sim r)$$
$$\wedge (p \vee \sim q \vee r) \wedge (p \vee q \vee r) \quad (6.2)$$

Reversing the order of these elementary disjuncts, we obtain the desired FCNF.

*Example 6.6.* Let us consider the function whose truth table is Table 6.3. A formula in FDNF for this function was obtained in Example 6.4. Similarly, an expression in FDNF for the negation of the function may be obtained:

$$(p \wedge q \wedge r) \vee (p \wedge q \wedge \sim r) \vee (p \wedge \sim q \wedge \sim r)$$
$$\vee (\sim p \wedge \sim q \wedge r) \quad (6.3)$$

Negating this and rearranging the terms, we find that the function may be expressed by (6.4).

$$(p \vee q \vee \sim r) \wedge (\sim p \vee q \vee r) \wedge (\sim p \vee \sim q \vee r)$$
$$\wedge (\sim p \vee \sim q \vee \sim r) \quad (6.4)$$

Shorter DNF and CNF for this function are

$$(p \wedge \sim q \wedge r) \vee (\sim p \wedge q) \vee (\sim p \wedge \sim r) \quad (6.5)$$

$$(p \vee q \vee \sim r) \wedge (\sim p \vee \sim q) \wedge (\sim p \vee r) \quad (6.6)$$

Although one use of these canonical forms is that of determining whether or not two formulas are logically equivalent, a more frequent use is that of finding a simpler formula logically equivalent to the given one.

The various minimization techniques discussed in the previous chapter may all be used here, as well as others which we have not discussed. These will lead to a minimal equivalent formula *in normal form,* but the reader is cautioned that if the use of other connectives is permitted the problem of finding a minimal formula is considerably more complex.

Perhaps the most important fact, for our purposes, is that these normal forms provide a means of generating a wff of an arbitrary number of variables and having a prescribed truth table. For we have given a procedure whereby the operations of conjunction, disjunction, and negation suffice to describe any truth table, no matter how long or complex. The formula which is obtained by this procedure may be quite lengthy, but it has the desired properties, and is subject to some simplification in general. We conclude this section with an example of a formula generated from the truth table of a four-variable function, both by use of the FDNF and by a more sophisticated analysis.

**Table 6.4**

A four-variable function

| $p$ | $q$ | $r$ | $s$ | $f(p, q, r, s)$ |
|-----|-----|-----|-----|-----------------|
| T | T | T | T | T |
| T | T | T | F | F |
| T | T | F | T | F |
| T | T | F | F | F |
| T | F | T | T | T |
| T | F | T | F | F |
| T | F | F | T | T |
| T | F | F | F | F |
| F | T | T | T | T |
| F | T | T | F | F |
| F | T | F | T | T |
| F | T | F | F | T |
| F | F | T | T | T |
| F | F | T | F | F |
| F | F | F | T | F |
| F | F | F | F | T |

*Example 6.7.* Consider Table 6.4. From this we may quickly develop an expression in FDNF for the given function, namely expression (6.7).

$$(p \wedge q \wedge r \wedge s) \vee (p \wedge \sim q \wedge r \wedge s) \vee (p \wedge \sim q \wedge \sim r \wedge s)$$
$$\vee (\sim p \wedge q \wedge r \wedge s) \vee (\sim p \wedge q \wedge \sim r \wedge s)$$
$$\vee (\sim p \wedge q \wedge \sim r \wedge \sim s) \vee (\sim p \wedge \sim q \wedge r \wedge s)$$
$$\vee (\sim p \wedge \sim q \wedge \sim r \wedge \sim s) \tag{6.7}$$

Simplifying transformations may now be applied to this. But rather than carry this out, we examine the truth table in detail. We note first that the 1st, 5th, 9th, and 13th rows all have the value T. These are exactly those lines for which $r$ and $s$ both have the value T. Hence the function may be represented by an expression "$(r \land s) \lor \cdots$." The remaining true lines in the table, namely the 7th, 11th, 12th, and 16th, are all lines for which $r$ has the value F. Thus the desired expression becomes "$(r \land s) \lor (\sim r \land (\ldots))$," and we have only to complete the missing part. To facilitate this we form Table 6.5, which is the subtable of Table 6.4 for which $r$ has

**Table 6.5**

Subtable of Table 6.4

| $r$ | $p$ | $s$ | $q$ | $f(p, q, r, s)$ |
|---|---|---|---|---|
| F | T | T | T | F |
| F | T | T | F | T |
| F | T | F | T | F |
| F | T | F | F | F |
| F | F | T | T | T |
| F | F | T | F | F |
| F | F | F | T | T |
| F | F | F | F | T |

the value F. In this subtable we have rearranged the variables in an order which is meant to be suggestive. In fact, looking at the table we see that if $p$ has the value F the value of the function corresponds to that of $s \supset q$, whereas if $p$ has the value T the function has the same value as $s \not\supset q$. Thus this subtable may be represented by the expression $\sim r \land (p \not\equiv (s \supset q))$, and hence the given function is represented by expression (6.8).

$$(r \land s) \lor (\sim r \land (p \not\equiv (s \supset q))) \tag{6.8}$$

By suitable transformations from Table 6.1, this becomes (6.9).

$$(r \lor (p \equiv (s \supset q))) \supset (r \land s) \tag{6.9}$$

EXERCISES

1–15. For each of the formulas in the exercises to Section 3 find the corresponding FDNF and FCNF, and write out equivalent formulas which in your opinion are as simple as possible.

16. Show that $(p \wedge q) \supset (\sim p \vee (p \mid \sim q)) \not\equiv \cdot \sim p \supset ((q \subset p) \downarrow \sim q)$ is logically equivalent to $p \downarrow q$. (See p. 60.)

17. Show that $(r \wedge s) \vee (\sim r \wedge (p \not\equiv (s \supset q)))$ eq $(r \vee (p \equiv (s \supset q))) \supset (r \wedge s)$. (See p. 66.)

## 7. "POLISH" NOTATION AND THE TREE OF A FORMULA

Throughout our work thus far, we have been bound by the necessity of using parentheses, or other devices such as an order of precedence and the dot system, in order to keep the syntax of our propositions clear. The need for this arises from the traditional use of an infix notation for binary operations: when we write A $\supset$ B, or A + B, or A $\triangle$ B, we must know how far to the left A extends and how far to the right B extends. Since we have been trained from childhood to use such a system of notation this does not cause great difficulties generally. But when we analyze the process of understanding the syntax of an expression, as we must in order to program a computer to "understand" the expression, it is evident that for a moderately complex expression this understanding involves a number of searches back and forth across the expression.

In 1951 the Polish logician Jan Łukasiewicz suggested that a prefix notation be used for all operations [3], so that we might write $\supset$ AB, + AB, or $\triangle$AB, and pointed out that consistent use of such a notation eliminates all need for punctuation. In this section we wish to examine such a notation and its relation to a certain graph, called the "tree" of a formula. Although the particular context we shall use is the propositional calculus, the notation is applicable wherever functions of $n$ variables ($n \geq 1$) are encountered or can be defined.

### Table 7.1

The relation between Polish and standard notations

| Polish | Standard |
| --- | --- |
| $Np$ | $\sim p$ |
| $Cpq$ | $p \supset q$ |
| $Apq$ | $p \vee q$ |
| $Kpq$ | $p \wedge q$ |
| $Epq$ | $p \equiv q$ |

A change in symbolism is in order at this point so that we do not make confusing double usage of certain symbols. The following symbolism will be used whenever we are writing in the prefix notation—propositions will be denoted by lower case letters: $p$, $q$, $r$, $s$; and operators or connectives will be denoted by capital letters: $N$, $C$, $A$, $K$, $E$. We shall also use lower case Greek letters to denote well-formed formulas (which must now be defined anew, since a different symbolism is involved). Table 7.1 conveys our intentions. Formally we define well-formed formulas recursively, as before.

**Definition 7.1.** A *well-formed formula* (of the propositional calculus, Polish notation) is a formula which may be obtained by finitely many applications of the following rules.

1. A propositional variable standing alone is a wff.
2. If $\alpha$ is a wff, then $N\alpha$ is a wff.
3. If $\alpha$ and $\beta$ are wffs, then $C\alpha\beta$, $A\alpha\beta$, $K\alpha\beta$, and $E\alpha\beta$ are wffs.

*Example 7.1.* (a) $(\sim p \supset r) \wedge (q \equiv p)$ in Polish notation is $KCNprEqp$.
(b) $(\sim q \wedge (p \equiv r)) \vee (\sim p \wedge q)$ becomes $AKNqEprKNpq$.
(c) $(\sim p \wedge (r \supset q)) \vee (p \wedge \sim q \wedge r)$ becomes either $AKNpCrqKpKNqr$ or $AKNpCrqKKpNqr$, depending on how one restores the parentheses in the last part of the statement. With our precedence conventions, the latter formula is the correct one.

Symbols may be defined for the remaining connectives such as $\not\equiv$ and $|$ if so desired. However, we shall have no need for these symbols.

Although this Polish notation may appear strange, it is an extremely easy notation to use because of its high directionality. We may best motivate our discussion by thinking of the logical connectives as being just that: symbols which connect together other strings of symbols. For example, $p$ and $q$ are two unrelated symbols, and the string $pq$ consists of two distinct parts, as would a run-on sentence. But $C$ as in $Cpq$, or $K$ as in $Kpq$, connects these two pieces and forms a single unified string. This provides the basis for our algorithm for well-formed formulas.

We assign *ranks* to our symbols as follows. A propositional variable has rank $+1$. (It alone forms a wff.) A binary connective ($C$, $A$, $K$, $E$) has rank $-1$. (It reduces the number of pieces in a string by joining two of them into one.) A singulary connective ($N$) has rank $0$. (It does not change the number of pieces: $p$ and $Np$ both have just one piece.) If we had ternary connectives they would have rank $-2$, and so forth.

### Algorithm for Well-Formed Formulas

Let $\mathbf{S} = s_1 s_2 \ldots s_{n-1} s_n$ be a formula in Polish notation, and let $r_i$ be the rank of $s_i$, $i = 1, 2, \ldots, n$. Beginning at the right-hand end of the formula, form the partial sums

$$\Sigma_n = r_n$$
$$\Sigma_{n-1} = r_{n-1} + r_n$$
$$\vdots$$
$$\Sigma_2 = r_2 + \cdots + r_{n-1} + r_n$$
$$\Sigma_1 = r_1 + r_2 + \cdots + r_{n-1} + r_n$$

Then **S** is a well-formed formula if and only if (1) each $\Sigma_i$ is positive ($\geq 1$), $i = 1, 2, \ldots, n$, and (2) $\Sigma_1 = 1$.*

*Example 7.2.* (a)

| Symbol string | K | C | N | p | r | E | q | p |
|---|---|---|---|---|---|---|---|---|
| Ranks | −1 | −1 | 0 | 1 | 1 | −1 | 1 | 1 |
| Partial sums | 1 | 2 | 3 | 3 | 2 | 1 | 2 | 1 |

Here, for example, $\Sigma_4 = 3$, indicating that at this point we have three unconnected pieces of formula, namely $p$, $r$, and $Eqp$. The formula is well-formed.

(b)

| Symbol string | A | K | N | p | C | r | q | K | p | K | N | q | r |
|---|---|---|---|---|---|---|---|---|---|---|---|---|---|
| Ranks | −1 | −1 | 0 | 1 | −1 | 1 | 1 | −1 | 1 | −1 | 0 | 1 | 1 |
| Partial sums | 1 | 2 | 3 | 3 | 2 | 3 | 2 | 1 | 2 | 1 | 2 | 2 | 1 |

This is also a wff.

(c)

| Symbol string | p | K | N | q | r |
|---|---|---|---|---|---|
| Partial sums | 2 | 1 | 2 | 2 | 1 |

This is not a wff, but has two parts, namely $p$ and $KNqr$.

(d)

| Symbol string | K | p | N | C | A | p | q |
|---|---|---|---|---|---|---|---|
| Partial sums | 0 | 1 | 0 | 0 | 1 | 2 | 1 |

This is not a wff as it lacks a second operand for $C$. Written in the standard notation this becomes $p \wedge \sim((p \vee q) \supset \quad )$.

In order to define the tree of a formula, we must introduce some concepts from graph theory.

**Definition 7.2.** A *graph* consists of a set $P$ of points (called *vertices*) and a set $L$ of lines (called *edges*) such that each line in $L$ has associated with it exactly two points in $P$, which are its vertices. If $a$ and $b$ are vertices of a

---

* Proof that this algorithm works is given in Burks, Warren, and Wright [1].

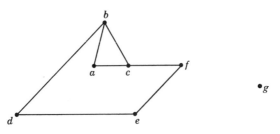

Figure 7.1. An example of a graph

graph, then a *chain from a to b* is a set $C = \{p_0, p_1, \ldots, p_n\} \subseteq P$ of vertices such that $a = p_0$, $b = p_n$, and for each $i = 1, \ldots, n$ there is an edge $e_i \in L$ whose vertices are $p_{i-1}$ and $p_i$. A graph is *connected* if there is a chain from any one of its vertices to any other. A *cycle* is a chain $\{p_0, p_1, \ldots, p_n\}$, $n \geq 1$, such that $p_0, p_1, \ldots, p_{n-1}$ are distinct and $p_n = p_0$. A connected graph with no cycles is called a *tree*.

*Example 7.3.* Figure 7.1 represents a graph whose vertices are $a$, $b$, $c$, $d$, $e$, $f$, and $g$, and whose edges are $ab$, $ac$, $bc$, $bd$, $cf$, $de$, and $ef$. There is a chain from $a$ to $f$: for example $\{a, b, c, f\}$, or $\{a, c, f\}$, or $\{a, b, d, e, d, b, c, f\}$. The graph is not connected, but would be if the point $g$ were deleted. It contains several cycles, as for example $\{a, b, c, a\}$, $\{b, d, e, f, c, b\}$, and $\{a, b, d, e, f, c, a\}$.

**Definition 7.3.** If $\alpha$ is a wff (in Polish notation) then the *tree of* $\alpha$ is the graph defined by the following process.

1. If $\alpha$ is a single propositional variable, then the tree of $\alpha$ is the graph with vertex $\alpha$ and no edges.

2. If $\alpha$ is of the form $N\beta$, then the tree of $\alpha$ is the graph whose vertices are $\{N\} \cup \{$vertices of $\beta\}$, and whose edges are the edges of $\beta$ together with an edge joining $N$ and the main connective of $\beta$ (or the vertex of $\beta$, if $\beta$ has but one vertex).

3. If $\alpha$ is of the form $X\beta\gamma$, where $X$ is one of the connectives $C$, $A$, $K$, and $E$, then the tree of $\alpha$ is the graph whose vertices are $\{X\} \cup \{$vertices of $\beta\}$ $\cup \{$vertices of $\gamma\}$, and whose edges are the edges of $\beta$ and of $\gamma$ together with an edge joining $X$ and the main connective of $\beta$, and an edge joining $X$ and the main connective of $\gamma$.

N

|

$T(\beta)$

(a)

X

$T(\beta)$     $T(\gamma)$

(b)

Figure 7.2. The tree of (a) $N\beta$, and (b) $X\beta\gamma$.

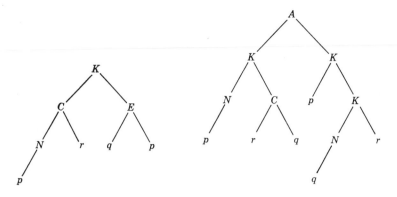

Figure 7.3. The tree of *KCNprEqp*.    Figure 7.4. The tree of *AKNpCrqKpKNqr*.

Symbolically, we may use the representation shown in Figure 7.2, where *T(β)* and *T(γ)* represent the trees of *β* and *γ* respectively.

*Example 7.4.* The trees of *KCNprEqp* and *AKNpCrqKpKNqr* are given in Figures 7.3 and 7.4 respectively.

Three comments are in order. First, the trees which we are using are in fact *labeled* trees. For example, the graph of Figure 7.5 could be the tree of many formulas. When the vertices are labeled, as in Figure 7.4, this becomes the tree of a particular formula. Second, because of the fact that *p* ⊃ *q* (*Cpq*) is not equivalent to *q* ⊃ *p* (*Cqp*), the left-to-right order of the vertices is important for our purposes. Third, the tree of a formula could just as well be developed from the standard infix notation. The present development was chosen because of the ease of constructing the tree.

If we know the symbols which occur on each level and the order in which they occur, it is easy to construct the tree. For example, if the first level contains the symbol *K*, the second the symbols *C*, *E*, the third *N*, *r*, *q*, *p*, and the fourth the symbol *p*, we may then construct the tree, shown in Figure 7.6. This information is easily obtained from the Polish notation for a propositional formula, by the following algorithm.

Figure 7.5. A tree.

1. Form the sum sequence $\Sigma_1, \ldots, \Sigma_n$ as above (p. 69). If the formula is well-formed, proceed.

2. Beginning with the left end, form two sequences $t_1, \ldots, t_n$ and $c_1, \ldots, c_n$ as follows.

    (a) $t_1 = 1, c_1 = 0$;

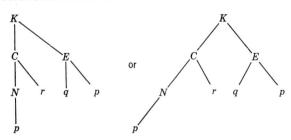

Figure 7.6. Construction of a tree with known symbol occurrences.

(b) for $i = 1, \ldots, n - 1$

if $\Sigma_{i+1} > \Sigma_i$, set $t_{i+1} = t_i + 1$, $c_{i+1} = 0$;

if $\Sigma_{i+1} = \Sigma_i$, set $t_{i+1} = t_i + 1$, $c_{i+1} = 1$;

if $\Sigma_{i+1} < \Sigma_i$, determine the last $k(\le i)$ such that $c_k = 0$; then set $t_{i+1} = t_k$, $c_{i+1} = 1$, $c_k = 1$.

3. If the given formula is well-formed and the above procedure has been properly followed, then at the end $c_1 = 0$, $c_i = 1$ ($i = 2, \ldots, n$), and $t_i$ is the level of the $i$th symbol in the tree, in its order of occurrence.

*Example 7.5.*

(a)

| Formula | $K$ | $C$ | $N$ | $p$ | $r$ | $E$ | $q$ | $p$ |
|---|---|---|---|---|---|---|---|---|
| $\Sigma$ | 1 | 2 | 3 | 3 | 2 | 1 | 2 | 1 |
| $t$ | 1 | 2 | 3 | 4 | 3 | 2 | 3 | 3 |
| $c$ | 0 | $\emptyset$ | $\emptyset$ | 1 | 1 | 1 | $\emptyset$ | 1 |
|  |  | 1 | 1 |  |  |  | 1 |  |

(b)

| Formula | $A$ | $K$ | $N$ | $p$ | $C$ | $r$ | $q$ | $K$ | $p$ | $K$ | $N$ | $q$ | $r$ |
|---|---|---|---|---|---|---|---|---|---|---|---|---|---|
| $\Sigma$ | 1 | 2 | 3 | 3 | 2 | 3 | 2 | 1 | 2 | 1 | 2 | 2 | 1 |
| $t$ | 1 | 2 | 3 | 4 | 3 | 4 | 4 | 2 | 3 | 3 | 4 | 5 | 4 |
| $c$ | 0 | $\emptyset$ | $\emptyset$ | 1 | 1 | $\emptyset$ | 1 | 1 | $\emptyset$ | 1 | $\emptyset$ | 1 | 1 |
|  |  | 1 | 1 |  |  | 1 |  |  | 1 |  | 1 |  |  |

(Compare Figure 7.3 and 7.4.)

Reviewing the above discussion, three facts become clear. First, the Polish notation for a formula is very closely related to the tree structure. In fact, beginning at the left end of the formula, if we examine the letters in sequence we are just tracing the tree of the formula, choosing at each step the left-most untraced branch. Second, the utility of the Polish notation for computation is evident: if we proceed from the right-hand end of a formula, then no operator or connective is encountered before its operands are known. Thus we need never delay a computation because the data are not known. Compare, for example, the computation of $(3 + 4)/(5 - (2 \times (1 + 3))) \times (2 - (4 \times 5))/(3 + (1 \times 2) - 4) + 2$ (from either end) with that of

$$AMDA34S5M2A13DS2M45SA3M1242$$

(from the right end), where $A$, $S$, $M$, $D$ denote add, subtract, multiply, and divide, respectively. Third, the lack of punctuation is dependent on having distinct symbols. Whenever we see "13" we must know whether this denotes the number thirteen, or the two numbers one and three. If the symbols lack distinctness, then punctuation must be reintroduced, at least to the extent of having some delimiter (perhaps a blank) to separate symbols.

We conclude this section with an example showing the development of a sparse truth table from a formula in Polish notation.

*Example 7.6.* Consider the formula

$$CEANpKpACpqrKsEprANKNpCpNKprCqKps$$

We wish to determine whether or not this formula is a tautology, and if not to perhaps develop a simpler equivalent formula. The work is given in Table 7.2, with the following notes.

**Table 7.2**

Truth table for Example 7.6

```
C E A N p K p A C p q  r K s E p r A N K N p C p N K p r C q K p s

1 2 3 4 4 3 4 3 4 5 4 3 2 3 2 3 2 1 2 2 3 3 2 3 2 2 3 2 1 2 1 2 1   (1)
1 2 3 4 5 4 5 5 6 7 7 6 3 4 4 5 5 2 3 4 5 6 5 6 6 7 8 8 3 4 4 5 5   (2)
F T                             F F T T F T         F T F           (3)
                                √ √ √ √   √           √   √         (4)
    T T F                                                           (5)
                            T T T F F                               (6)

                            p q r s   f(p, q, r, s)

                            F T F T       F                         (7)
                            Otherwise     T
```

1. This is the test for a wff.
2. This develops the tree structure.
3. The main connective is $C$, which has the value F in only one case. Working this out we find that $p$ has the value F, $q$ has the value T.

4. Using these values for $p$ and $q$ we find that the checked connectives have their proper values regardless of the values of $r$ and $s$.

5. Using the value of $p$ in the left-hand side, we find that the third symbol, $A$ has the value T.

6. Hence, since the second symbol, $E$, has the value T from (3), the $K$ at level 3 must also have the value T. This leads us to the values for $r$ and $s$.

7. These two lines, then, constitute the sparse truth table for this function. Note that equivalent formulas are $NKNpKqKNrs$ and $ApANqArNs$.

EXERCISES

1. Determine whether or not the following are wffs.
   (a) $CpKEpqANqNCpEqr$
   (b) $CAKEpNqNrNCAKEprNqNpr$
   (c) $ApCqErKpNq$
   (d) $EEpErENpNrpp$
   (e) $EANpKpNCpqKpNApqCKApqp$

2–11. Rewrite the formulas of Exercises 1–10 for Section 3 in Polish notation.

   In each of the following problems, (a) develop the tree structure, (b) develop the sparse truth table directly from the Polish notation formula, and (c) rewrite the formula in standard notation.

12. $EpEqErEpNq$
13. $ApCKNpNqKpEqr$
14. $CNNpAKNpNANCEpqrArNKrpq$
15. $ACANKAEEpqrspqrs$
16. $KANKAKEEppppqqqq$
17. $ArCANsqCAKEpqrps$

## 8. MINIMAL SETS OF CONNECTIVES

The question which concerns us in this section is that of how few connectives we really need to specify in order to develop our logic. Although this is of theoretical interest, the answer to the question also has practical implications. For if we are designing a logic machine we would like to standardize the basic elements of the machine. There is not much point in using ten types of basic elements if two or three will do just as well. On the other hand, there is a certain loss in using fewer elements: the logical expressions, and hence the circuitry, for representing particular functions may become more complex. Thus in practical design problems we must carefully balance the number of different types of basic elements against the circuitry necessary to use these elements.

Since any function can be represented by an expression in disjunctive normal form, it is clear that at most we need the connectives conjunction, disjunction, and negation. Moreover, De Morgan's laws allow us to express disjunction in terms of conjunction and negation ($p \lor q$ eq $\sim(\sim p \land \sim q)$), or conjunction in terms of disjunction and negation. Thus all we really need to have at our disposal is negation, and either conjunction or disjunction. Can we do better?

Let us work with disjunction and negation. Since negation is an operation involving only one operand, we cannot eliminate the disjunction: there would be no way to express functions of more than one variable. If we could represent negation by a disjunctive expression, that expression would involve only one propositional variable since negation is a singulary function. Thus $\sim p$ would have to be expressed by $p \lor p$, $(p \lor p) \lor p$, $(p \lor p) \lor (p \lor p)$, or some other such expression. But we can quickly see that any such expression is logically equivalent to $p$. Hence there is no disjunctive expression representing $\sim p$, and thus we must retain negation. Thus any logical function can be expressed in terms of disjunction and negation, and there are functions which cannot be expressed in terms of just one of these. We say that disjunction and negation form a *minimal set of connectives* for the propositional calculus.

Similarly, conjunction and negation form a minimal set of connectives.

It is possible to select other minimal sets of connectives also. In fact, the Sheffer's stroke by itself forms a minimal set of connectives, as does Peirce's arrow. To show that the stroke constitutes a minimal set we need merely to show that both negation and disjunction (or conjunction) are expressible in terms of the stroke, for then any function could be so expressed. By checking the truth tables the reader can verify that

$$\sim p \text{ eq } p \mid p, \quad p \lor q \text{ eq } (p \mid p) \mid (q \mid q), \quad \text{and} \quad p \land q \text{ eq } (p \mid q) \mid (p \mid q)$$

### EXERCISES

1. Verify the equivalences in the last statement of this section.
2. Show that the following are minimal sets of connectives.
   (a) $\{\sim, \supset\}$
   (b) $\{\downarrow\}$
3. Show that $\{\sim, \equiv\}$ is not a minimal set of connectives.

## 9. AN AXIOMATIC APPROACH TO LOGIC

At the beginning of this chapter we commented on the propositional calculus being a means of calculating with propositions. Thus far our calculations have been limited for the most part to computing truth tables and finding logically equivalent propositions. In this section we wish to approach the subject differently and to show how new expressions can be derived from given ones.

The approach which we now take is a formal one, involving only a procedure for forming new strings of symbols from given strings. For the moment the concept of truth does not enter the picture, although we shall see later that the strings which we derive from our axioms are in fact the tautologies.

The symbol strings which we shall use will be those formed from the propositional variables, connectives, and parentheses as we have been doing, except that the only "legitimate" connectives will be the negation and conditional. Expressions involving other connectives will be used only as abbreviations for expressions in these two connectives. (See Exercise 2a of Section 8.) With that proviso, the rules for defining well-formed formulas are those given in Section 2.

The process of generating strings begins with the selection of a non-empty set of wffs, called the set of *axioms*. These constitute the basis for generation: the strings which we generate will be constructed out of the axioms. Thus the choice of the set of axioms materially affects the set of strings which are generated. We choose a set of three axioms:

**Axiom 1.** $p \supset (q \supset p)$

**Axiom 2.** $(p \supset (q \supset r)) \supset ((p \supset q) \supset (p \supset r))$

**Axiom 3.** $(\sim p \supset \sim q) \supset ((\sim p \supset q) \supset p)$

Once a set of axioms has been chosen, the next step is to specify a set of *rules of inference* by which the new strings are to be generated. We choose two rules of inference:

#### Modus Ponens

Given strings A and A $\supset$ B, it is permissible to generate string B (see Example 5.1).

#### Substitution

Given string A, well-formed formula B, and a propositional variable, say $p$, it is permissible to generate a string C by substituting B for each occurrence of $p$ in A.

In some presentations the use of the substitution rule is avoided by choosing an infinite set of axioms, which are represented by *axiom schemata*. For example, in place of our Axiom 1 we would have Axiom Schema 1: A $\supset$ (B $\supset$ A), where A and B denote wffs as usual. This axiom schema would then denote the set of all such formulas, which would be a subset of the set of axioms. In our presentation the formulas in this set must be generated by substitution in the given axiom.

**Definition 9.1.** A wff A is a *theorem* if there exists a finite sequence of wffs $A_1, A_2, \ldots, A_n$ such that (1) each $A_i$ is either an axiom or is derived from preceding formulas in the list by one of the rules of inference, and (2) $A_n$ is the wff A. Such a sequence is called a *proof*, in particular a *proof of A*. If A is a theorem, we write $\vdash A$. In particular, an axiom is a theorem since there exists a one line proof of it.

Notice that this definition gives us a precise formulation of the concepts of theorem and proof *within the theory*. As mentioned in Section 2, these concepts are less well-defined within the metatheory. We illustrate the process of generating strings by proving three theorems.

**Theorem 9.1.** $\vdash p \supset p$.

*Proof.*  (1) $p \supset (q \supset r) \supset .(p \supset q) \supset (p \supset r)$     [Axiom 2]

(2) $p \supset ((p \supset p) \supset p)$
$\supset .(p \supset (p \supset p)) \supset (p \supset p)$     [Substitution (1)]

(3) $p \supset (q \supset p)$     [Axiom 1]

(4) $p \supset ((p \supset p) \supset p)$     [Substitution (3)]

(5) $p \supset (p \supset p) \supset .p \supset p$     [Modus Ponens (4), (2)]

(6) $p \supset (p \supset p)$     [Substitution (3)]

(7) $p \supset p$     [Modus Ponens (6), (5)]

(Step (2) actually includes two substitutions.)

**Theorem 9.2.** $\vdash (\sim p \supset p) \supset p$.

*Proof.*  (1) $(\sim p \supset \sim q) \supset .(\sim p \supset q) \supset p$     [Axiom 3]

(2) $(\sim p \supset \sim p) \supset .(\sim p \supset p) \supset p$     [Substitution (1)]

(3) $p \supset p$     [Theorem 9.1]

(4) $\sim p \supset \sim p$     [Substitution (3)]

(5) $(\sim p \supset p) \supset p$     [Modus Ponens (4), (2)]

Notice that what we have given here is not a proof as we have defined it, but rather an abbreviation for one. To obtain a legitimate proof one should substitute for (3) a proof of that statement (for example, that given in Theorem 9.1).

**Theorem 9.3.** $\vdash (q \supset r) \supset .(p \supset q) \supset (p \supset r)$.

*Proof.*  (1) $p \supset (q \supset r) \supset .(p \supset q) \supset (p \supset r)$     [Axiom 2]

(2) $p \supset (q \supset p)$     [Axiom 1]

(3) $(p \supset (q \supset r)) \supset ((p \supset q) \supset (p \supset r))$
$\supset .(q \supset r) \supset ((p \supset (q \supset r))$
$\supset ((p \supset q) \supset (p \supset r)))$     [Substitution (2)]

(4) $q \supset r \supset .(p \supset (q \supset r))$
$\qquad \supset ((p \supset q) \supset (p \supset r))$      [Modus Ponens (1), (3)]

(5) $(q \supset r) \supset (p \supset (q \supset r))$      [Substitution (2)]

(6) $(q \supset r) \supset ((p \supset (q \supset r))$
$\qquad \supset ((p \supset q) \supset (p \supset r)))$
$\qquad \supset .((q \supset r) \supset (p \supset (q \supset r)))$
$\qquad \supset ((q \supset r) \supset ((p \supset q) \supset (p \supset r)))$      [Substitution (1)]

(7) $(q \supset r) \supset (p \supset (q \supset r))$
$\qquad \supset .(q \supset r) \supset ((p \supset q) \supset (p \supset r))$      [Modus Ponens (4), (6)]

(8) $(q \supset r) \supset .(p \supset q) \supset (p \supset r)$      [Modus Ponens (5), (7)]

One question which arises in studying an axiom system such as this one is its relation to the truth functions which we have examined earlier. It would be nice to know that everything which is provable is true and vice versa. For the propositional calculus this is indeed the case. As the reader can easily verify, all of our axioms are tautologies; by Example 5.1 modus ponens preserves tautologies. That is, if A and A $\supset$ B are tautologies, then so also is B. Furthermore, since any substitution is done at *all* occurrences of a propositional variable, tautologies are also preserved under this rule of inference. Thus any theorem is a tautology. The converse statement—that any tautology is a theorem—is also true, but the proof of it is beyond the scope of this book. This latter statement is known as the *completeness theorem* for the propositional calculus.*

Another question which arises is what would happen if the axiom system were modified or the rules of inference changed. If we add a statement that is a theorem to a given axiom system, then the system is not really affected except that it becomes less elegant. In any instance when the new statement is used as an axiom, a proof of it from the original axioms may be inserted. Thus the set of theorems (that is, provable statements) remains unchanged.

Properly done, quite extensive modifications can be made in the set of axioms and the rules of inference without affecting the set of theorems. In fact, several different systems have been proposed, each strongly defended as being in some way preferable.†

However, the addition of an arbitrary statement to the axiom set may result in chaos: often any well-formed formula becomes provable. Nevertheless such an addition can be of great use at times. Thus, neither $p \wedge q$

---

* For a proof of the completeness theorem, see Mendelson [4], or almost any other of the standard logic texts.
† See, for example, Mendelson [4] and Church [2].

nor $p \lor \sim r$ are theorems (since they are not tautologies); but the addition of $p \land q$ *as a hypothesis* allows us to deduce $p \lor \sim r$ as a consequence, since $(p \land q) \supset (p \lor \sim r)$ is a tautology and hence a theorem. Thus we can state that whenever $p \land q$ is true then $p \lor \sim r$ is also true.

**Definition 9.2.** Let $\{A_1, A_2, \ldots, A_n, B\}$ be a finite set of wffs. We say that B is a *consequence* of, or is *deducible* from, $A_1, A_2, \ldots, A_n$ (written $A_1, A_2, \ldots, A_n \vdash B$) if there exists a finite sequence of wffs, $P_1, P_2, \ldots, P_k$, such that (1) each $P_i$ is either an axiom, or one of the *hypotheses* $A_1, \ldots, A_n$, or is derived from preceding formulas in the sequence by one of the rules of inference, (2) substitution is never applied to any variable occurring in the hypotheses, and (3) $P_k$ is the wff B. The sequence $P_1, \ldots, P_k$ is called a *deduction* of B from hypotheses $A_1, \ldots, A_n$. If B is a theorem, it is a consequence of the empty set of hypotheses, and we write $\vdash B$.

**The Deduction Theorem.** Let $\Gamma$ be a set of wffs, and let A and B be wffs. If $\Gamma, A \vdash B$, then $\Gamma \vdash A \supset B$. In particular, if $A \vdash B$, then $\vdash A \supset B$.

This theorem was originally proven by Herbrand in 1930, and a proof of it may be found in any standard logic text. Basically the idea of the proof is to look at a deduction of B from $\Gamma$, A, and to replace each line $P_i$ of that deduction by the line $A \supset P_i$. What results is no longer a deduction, but by the interpolation of suitable additional lines becomes a deduction of $A \supset B$ from the hypotheses $\Gamma$. The Deduction Theorem is an invaluable tool, for it enables one to write down quickly proofs of theorems which might otherwise be quite involved.

*Example 9.1.* As an example, we reprove Theorem 9.3, taking as hypotheses $q \supset r$, $p \supset q$, and $p$.

(1) $p$           [Hypothesis]
(2) $p \supset q$    [Hypothesis]
(3) $q$          [Modus Ponens (1), (2)]
(4) $q \supset r$    [Hypothesis]
(5) $r$          [Modus Ponens (3), (4)]

Thus we have a deduction $q \supset r$, $p \supset q$, $p \vdash r$. Applying the Deduction Theorem three times, we obtain successively

$$q \supset r, p \supset q \vdash p \supset r$$
$$q \supset r \vdash (p \supset q) \supset (p \supset r)$$
$$\vdash (q \supset r) \supset ((p \supset q) \supset (p \supset r))$$

What we have now is not a proof of Theorem 9.3 as we have defined "proof" within our theory. But the techniques used in proving the Deduction Theorem are constructive, and by using these we can transform the deduction that we have written down into a valid proof of the theorem. As this is a tedious mechanical task it is generally not carried out.

Thus we have the counterpart here to the relationship between logical consequence and tautology mentioned in Section 5. This correspondence and the correspondence between tautologies and theorems make much of the work in the propositional calculus trivial. Is a given wff a theorem? To find out we need merely to check its truth table—a mechanical process. If it is a theorem, what about a proof? First replace the problem by that of finding a suitable deduction. Once this has been found another mechanical process will transform the deduction into the desired proof. The only hard part left is that of finding a suitable deduction; for this there are few known guides between brute force on the one end and a highly developed intuition on the other.

EXERCISES

In each of the problems, give a correct proof or deduction.

1. $p \supset q, q \supset r \vdash p \supset r$
2. $p \supset (q \supset r), q \vdash p \supset r$
3. $\vdash (\sim p \supset \sim \sim p) \supset p$
4. $\vdash \sim \sim p \supset p$
5. $\vdash (\sim \sim \sim p \supset p) \supset \sim \sim p$
6. $\vdash p \supset \sim \sim p$
7. $\sim p \vdash (\sim q \supset p) \supset q$
8. $\vdash \sim p \supset (p \supset q)$
9. $\vdash p \wedge q \supset p$
10. $\vdash p \wedge q \supset q$
11. $\vdash (\sim q \supset \sim p) \supset (p \supset q)$
12. $p \supset q \vdash \sim \sim p \supset q$
13. $p \supset q \vdash \sim \sim p \supset \sim \sim q$
14. $\vdash (p \supset q) \supset (\sim q \supset \sim p)$
15. $(p \supset q) \supset q \vdash \sim q \supset \sim (p \supset q)$
16. $\vdash p \supset (\sim q \supset \sim (p \supset q))$
17. $\vdash p \supset (q \supset (p \wedge q))$
18. $\sim p \supset q \vdash (\sim q \supset \sim p) \supset q$
19. $\vdash (p \supset q) \supset ((\sim p \supset q) \supset q)$

**References**

1. Burks, A. W., D. W. Warren, and J. B. Wright. "An analysis of a logical machine using parenthesis-free notation," *Mathematical Tables and Other Aids to Computation*, Vol. 8, No. 46, April 1954, 53–57.
2. Church, A. *Introduction to Mathematical Logic*, Vol. I. Princeton University Press, Princeton, New Jersey, 1956.
3. Łukasiewicz, Jan. *Aristotle's Syllogistic from the Standpoint of Modern Formal Logic*. Oxford University Press, New York, 1951.
4. Mendelson, E. *Introduction to Mathematical Logic*. D. Van Nostrand Company, Princeton, New Jersey, 1964.

5. Rosenbloom, P. C. *The Elements of Mathematical Logic.* Dover Publications, New York, 1950.
6. Stoll, R. R. *Set Theory and Logic.* W. H. Freeman and Company, San Francisco, 1963.
7. Suppes, P. *Introduction to Logic.* D. Van Nostrand Company, Princeton, New Jersey, 1957.
8. *More Problematical Recreations.* Litton Industries, Beverley Hills, California (undated).

# 4. A View of Binary Vectors

In this chapter we wish to examine a particular structure and see how it may be interpreted in various ways. This structure is a *vector* or ordered $n$-tuple of 0's and 1's. We shall consider the four following vectors throughout as our examples.

$$\mathbf{V}_1 = \langle 0, 1, 1, 0, 1, 0 \rangle \quad \text{or} \quad 0\ 1\ 1\ 0\ 1\ 0$$
$$\mathbf{V}_2 = \langle 1, 0, 1, 1, 0, 1 \rangle \quad \text{or} \quad 1\ 0\ 1\ 1\ 0\ 1$$
$$\mathbf{V}_3 = \langle 1, 0, 0, 1, 1, 0 \rangle \quad \text{or} \quad 1\ 0\ 0\ 1\ 1\ 0$$
$$\mathbf{V}_4 = \langle 0, 1, 1, 0, 0, 1 \rangle \quad \text{or} \quad 0\ 1\ 1\ 0\ 0\ 1$$

## 1. SETS

We may use vectors to represent sets by allowing each component or element of the vector to denote a particular element of the universe. We then use a "1" to denote the fact that the element represented by a given component is in the set we are considering, and a "0" to indicate that it is not. For example, if the universe is $U = \{1, 2, 3, 4, 5, 6\}$ and we let the $i$th component denote the element $i$, then $\mathbf{V}_1$ represents the set $A = \{2, 3, 5\}$, $\mathbf{V}_2$ represents $B = \{1, 3, 4, 6\}$, and so forth.

The various set operations are now easily accomplished by comparing the vector representations for the presence or absence of 1's. Set union, for example, requires the presence of a 1 in a given position in either vector, whereas set intersection demands the presence of a 1 in a given position of both vectors.

*Example 1.1.*

(a)      $A$:   0 1 1 0 1 0
            $B$:   1 0 1 1 0 1
     $A \cup B$:   1 1 1 1 1 1      (1 in either $A$ or $B$)    that is, $A \cup B = U$

(b)      $A$:   0 1 1 0 1 0
            $B$:   1 0 1 1 0 1
     $A \cap B$:   0 0 1 0 0 0      (1 in both $A$ and $B$)    that is, $A \cap B = \{3\}$

(c)      $A$:   0 1 1 0 1 0
            $B$:   1 0 1 1 0 1
     $A \triangle B$:   1 1 0 1 1 1      (1 in exactly one of $A$ and $B$)    that is,
                 $A \triangle B = \{1, 2, 4, 5, 6\}$

(d) $C$:  1 0 0 1 1 0     that is, $C = \{1, 4, 5\}$
  $\bar{C}$:  0 1 1 0 0 1     (interchange 0 and 1)   that is, $\bar{C} = \{2, 3, 6\}$

EXERCISE

1. Let the universe consist of the digits 0 through 9.
   (a) Write a vector to describe each of the following sets.
      $A = \{1, 2, 5\}$
      $B = \{1, 2, 5, 7, 8\}$
      $C = \{2, 5, 6, 7\}$
      $D = \{2, 3, 4\}$
      $E = \{1, 6, 7, 9\}$
      $F = \varnothing$
      $G = U$
   (b) Use these vectors to compute the following sets.
      $R = A \cap (\overline{B \cap C})$
      $S = A \cup (B \triangle (\bar{D} \triangle E))$
      $T = B \cup (D \cap \bar{E})$

## 2. LOGIC

Let us now suppose that we have a set of objects coded according to the presence or absence of a number of properties—in our examples, six properties. We may use the code of a "1" to indicate the presence of a property, and a "0" to indicate the absence of this property. In other words, the 1 and 0 now take the place of the letters T and F respectively for the sentence "This object has property $p_i$." For example, in scheduling problems the objects might be people, and the properties might be training to operate certain machines, availability at certain times, willingness to travel, etc. In information retrieval work, as applied to a document, the 1 and 0 would indicate the presence or absence of certain keywords in the document; as applied to a request, the same 1 or 0 would indicate an interest or a lack of interest in those keywords. We may then determine various combinations of these properties by applying the logical operations to them.

*Example 2.1.* The objects are people, the properties are available times. When can $\mathbf{V}_2$ and $\mathbf{V}_3$ meet?

$$\mathbf{V}_2\text{:}\quad 1\ 0\ 1\ 1\ 0\ 1$$
$$\mathbf{V}_3\text{:}\quad 1\ 0\ 0\ 1\ 1\ 0$$
$$\mathbf{V}_2 \wedge \mathbf{V}_3\text{:}\quad 1\ 0\ 0\ 1\ 0\ 0$$

Thus $\mathbf{V}_2$ and $\mathbf{V}_3$ can meet at either time $t_1$ or time $t_4$.

*Example 2.2.* The objects are watchmen, the properties are times on duty. When is at least one of the watchmen $V_1$ or $V_3$ on duty?

$$\begin{array}{ll} V_1: & 0\ 1\ 1\ 0\ 1\ 0 \\ V_3: & 1\ 0\ 0\ 1\ 1\ 0 \\ V_1 \vee V_3: & 1\ 1\ 1\ 1\ 1\ 0 \end{array}$$

One of the watchmen is on duty at all times but $t_6$. What combinations of watchmen would provide complete time coverage?

$$\begin{array}{ll} V_1: & 0\ 1\ 1\ 0\ 1\ 0 \\ V_2: & 1\ 0\ 1\ 1\ 0\ 1 \\ V_1 \vee V_2: & 1\ 1\ 1\ 1\ 1\ 1 \end{array}$$

Thus together $V_1$ and $V_2$ would provide complete time coverage. Similarly, $V_3$ and $V_4$ provide complete coverage.

*Example 2.3.* The objects are documents, the properties keywords. We are looking for all documents having keywords $p_1$ and $p_4$. That is, we want a 1 in the first and fourth positions of the vector, but do not care what the other positions contain. This can be checked by using the conditional. Form the property vector $P = 1\ 0\ 0\ 1\ 0\ 0$, and check:

$$\begin{array}{ll} P: & 1\ 0\ 0\ 1\ 0\ 0 \\ V_1: & 0\ 1\ 1\ 0\ 1\ 0 \\ P \supset V_1: & 0\ 1\ 1\ 0\ 1\ 1 \end{array} \qquad \begin{array}{ll} P: & 1\ 0\ 0\ 1\ 0\ 0 \\ V_2: & 1\ 0\ 1\ 1\ 0\ 1 \\ P \supset V_2: & 1\ 1\ 1\ 1\ 1\ 1 \end{array}$$

$$\begin{array}{ll} P: & 1\ 0\ 0\ 1\ 0\ 0 \\ V_3: & 1\ 0\ 0\ 1\ 1\ 0 \\ P \supset V_3: & 1\ 1\ 1\ 1\ 1\ 1 \end{array} \qquad \begin{array}{ll} P: & 1\ 0\ 0\ 1\ 0\ 0 \\ V_4: & 0\ 1\ 1\ 0\ 0\ 1 \\ P \supset V_4: & 0\ 1\ 1\ 0\ 1\ 1 \end{array}$$

Those with ones in each position ($V_2$ and $V_3$) are the desired documents. Now suppose that we want those documents having keywords $p_1$ and $p_4$, but not $p_6$. Form property vector $Q = 1\ 1\ 1\ 1\ 1\ 0\ (\sim p_6)$, and check against $V_2$ and $V_3$, again using the conditional:

$$\begin{array}{ll} V_2: & 1\ 0\ 1\ 1\ 0\ 1 \\ Q: & 1\ 1\ 1\ 1\ 1\ 0 \\ V_2 \supset Q: & 1\ 1\ 1\ 1\ 1\ 0 \end{array} \qquad \begin{array}{ll} V_3: & 1\ 0\ 0\ 1\ 1\ 0 \\ Q: & 1\ 1\ 1\ 1\ 1\ 0 \\ V_3 \supset Q: & 1\ 1\ 1\ 1\ 1\ 1 \end{array}$$

Thus $V_3$ is the document for which we are searching.

## EXERCISES

1. Suppose that a library has documents characterized by keywords as follows:

   A:   city government, police protection, taxes

   B:   Congress, politics, taxes, welfare programs

C:  city government, police protection, street lighting
D:  civil rights, city government, politics
E:  city government, economics, politics, welfare programs
F:  Congress, education, taxes, welfare programs
G:  economics, street lighting
H:  city government, police protection, politics
 I:  Congress, economics, welfare programs
 J:  civil rights, city government, Congress
K:  street lighting, taxes
L:  city government, police protection, politics
M:  economics, education, welfare programs
N:  Congress, economics, education, taxes
O:  civil rights, Congress, economics

Make up vectors describing each of these documents, and use these vectors to answer the following requests:

α:  civil rights, and either city government or police protection
β:  education or welfare programs, but if the latter only if either taxes or Congress is included
γ:  Congress and police protection

## 3. NUMBERS

The ordinary numerical notation which we use is a positional or vector notation with a base of ten. Thus, 1473 denotes the number $1 \times 10^3 + 4 \times 10^2 + 7 \times 10^1 + 3 \times 10^0$. We may use the same type of notation with any positive integer base $r$, using only the digits $0, 1, \ldots, r - 1$. For example, if we used base 8, then 1473 would denote the number

$$1 \times 8^3 + 4 \times 8^2 + 7 \times 8^1 + 3 \times 8^0$$

In particular, we are interested in using the base 2. In this instance the only digits used are 0 and 1, and the numeral 100110, for example, would be $1 \times 2^5 + 0 \times 2^4 + 0 \times 2^3 + 1 \times 2^2 + 1 \times 2^1 + 0 \times 2^0$, or 38 in decimal notation. Similarly, $V_1$ would represent the decimal numeral 26, $V_2$ would represent 45, and $V_4$ would be 25. Notice that we are using the convention of allowing the *binary numerals* to begin with a zero, so that we may write three either as 11 ($= 1 \times 2^1 + 1 \times 2^0$) or as 011 ($= 0 \times 2^2 + 1 \times 2^1 + 1 \times 2^0$). This allows us to use vectors of fixed length, and is a common convention among computer users.

Arithmetic in the binary system is exceedingly simple:

$$0 + 0 = 0$$
$$0 + 1 = 1$$
$$1 + 0 = 1$$
$$1 + 1 = 10 \qquad \text{(that is, } 1 + 1 = 2, \text{ decimal)}$$

$$0 \times 0 = 0$$
$$0 \times 1 = 0$$
$$1 \times 0 = 0$$
$$1 \times 1 = 1$$

*Example 3.1.*

| | | |
|---|---|---|
| $V_1$: | 0 1 1 0 1 0 | 26 |
| $V_2$: | 1 0 1 1 0 1 | +45 |
| $V_1 + V_2$: | 1 0 0 0 1 1 1 | 71 |

$V_1$:　　　　　　　　0 1 1 0 1 0
$V_4$:　　　　　　　　0 1 1 0 0 1

　　　　　　　　　　0 1 1 0 1 0
　　　　　　　0 1 1 0 1 0 0 0
　　　　0 1 1 0 1 0

$V_1 \times V_4$:　　1 0 1 0 0 0 1 0 1 0
$(26 \times 25 = 650)$

This arithmetic is also intimately connected with our logical system, as we can see by examining addition in more detail. When adding two digits, we obtain a two-digit result, involving the *sum* and *carry* digits as shown in Table 3.1.

**Table 3.1**

Sum and carry digits in binary addition

| $a$ | $b$ | Carry | Sum |
|---|---|---|---|
| 0 | 0 | 0 | 0 |
| 0 | 1 | 0 | 1 |
| 1 | 0 | 0 | 1 |
| 1 | 1 | 1 | 0 |

Now if we think of the 0 and 1 as F and T, as before, we see that the carry digit of $a + b$ is given by $a \wedge b$, and the sum digit by $a \not\equiv b$. Thus addition may be performed by using the logical operations of conjunction and nonequivalence together with a shift operation which moves the carry digit over to the next vector position.

*Example 3.2.* Add $V_1$ and $V_2$.

$$
\begin{array}{lllllll}
V_1: & 0 & 1 & 1 & 0 & 1 & 0 \\
V_2: & 1 & 0 & 1 & 1 & 0 & 1 \\
\text{Sum}_1: & 1 & 1 & 0 & 1 & 1 & 1 \quad \text{Carry}_1: \quad 0\ 0\ 1\ 0\ 0\ 0
\end{array}
$$

Shift the carry over one position (0 1 0 0 0 0) and add again:

$$
\begin{array}{lllllll}
\text{Sum}_1: & 1 & 1 & 0 & 1 & 1 & 1 \\
\text{(Shifted) Carry}_1: & 0 & 1 & 0 & 0 & 0 & 0 \\
\text{Sum}_2: & 1 & 0 & 0 & 1 & 1 & 1 \quad \text{Carry}_2: \quad 0\ 1\ 0\ 0\ 0\ 0
\end{array}
$$

Repeat until there is no carry:

$$
\begin{array}{lllllll}
\text{Sum}_2: & 1 & 0 & 0 & 1 & 1 & 1 \\
\text{Carry}_2: & 1 & 0 & 0 & 0 & 0 & 0 \\
\text{Sum}_3: & 0 & 0 & 0 & 1 & 1 & 1 \quad \text{Carry}_3: \quad 1\ 0\ 0\ 0\ 0\ 0
\end{array}
$$

$$
\begin{array}{lllllll}
\text{Sum}_3: & 0 & 0 & 0 & 0 & 1 & 1 & 1 \\
\text{Carry}_3: & 1 & 0 & 0 & 0 & 0 & 0 & 0 \\
\text{Sum}_4: & 1 & 0 & 0 & 0 & 1 & 1 & 1 \quad \text{Carry}_4: \quad 0\ 0\ 0\ 0\ 0\ 0\ 0
\end{array}
$$

Thus $\text{Sum}_4$ is the desired result.

Similarly, the other arithmetic operations may be related to our basic logic. Thus we see that one may take a simple vector notation and interpret it in a number of ways; and that, in fact, whichever interpretation we choose to make, the basic operations are closely related. Hence one may build a device utilizing these basic operations (for example, negation or complementation, conjunction, and a shift), and allow the potential user to interpret the results of any computation according to his needs.

## EXERCISES

1. Convert these binary numerals to decimal.
   - (a) 10011011
   - (b) 11100
   - (c) 0010101
   - (d) 1010011000111
   - (e) 1001010
   - (f) 00000010
2. Convert these decimal numerals to binary.
   - (a) 17
   - (b) 1492
   - (c) 10101
   - (d) 12345
   - (e) 24
   - (f) 692
3. Relate binary multiplication to the logical operations, and set up a routine to perform this multiplication using only logical operations and a shift, assuming that we wish to multiply two numbers having three binary digits each.

# 5. Algorithms and Computing Machines

## 1. ALGORITHMS: METHODS OF SOLVING PROBLEMS

What sorts of problems might one solve with a simple logic machine such as mentioned in the last chapter? To put it another way, what is required of a problem in order that we may solve it; what is required of both the problem and the machine in order that we may describe to the machine the problem and a method for solving it, and expect the machine to solve the problem? In this chapter we will discuss these questions, and describe in more detail some problem-solving devices and their limitations. The logic which we have developed will serve both directly for examples, and indirectly as a base for our discussion.

We begin with certain generalizations which are applicable to both man and machine. First, if a method is not known for solving a given problem, then there must be a more or less thorough understanding of the problem in order to arrive at a solution. Very little is presently known about how to impart such understanding to a machine. Hence, *we will assume for our work that we know a method for solving a given problem.*

Second, if a "good" method is known for solving a problem, then the problem need not be understood at all. For example, a clerk can keep the account books for a small firm without understanding the firm's finances, provided that he knows arithmetic, is told exactly what to do, and does exactly as he is told. However, in such a situation *we must specify exactly and completely the method of solution in a language which the device can interpret.*

Third, problems may be grouped into sets of problems which are in some sense similar. *Generally a method for solving all problems of a given set is deemed superior to a method for solving only one problem of the set.*

Fourth, *there are criteria for determining how "good" a solution method is other than whether or not it works, or how generally applicable the method is.* These criteria involve such things as efficiency, elegance, and speed, and are frequently very ill-defined.

Let us turn now to the problem of defining more exactly what is meant by a method of solution of a problem. If we are given a problem, there may either be a solution (find $x$ such that $2x = 6$) or there may not be a solution (find $x$ such that $0x = 6$). In some cases it may not be clear whether the problem has a solution; indeed, in such areas as the social sciences it is often difficult to get people to agree as to what constitutes a solution. Certainly if a problem has a well-defined solution we would insist that our "method of solution" find it; but in case there is no solution, our "method of solution" may or may not recognize that fact.

The method of solution must, of course, be compatible with the device used to solve the problem. That is, it must be stated in a "language" which the device can comprehend, and must be a procedure which the device is capable of executing. Consider, for example, the problem of extracting a gumball from a gumball machine. The device may be considered to be the gumball machine. In this case the "language" is a very simple one: it consists of a penny and a push on the lever of the machine. The machine then "understands" that a penny followed by a push means "emit one gumball," and it reacts accordingly. On the other hand, we may consider the device to consist of a small boy and a gumball machine, in which case the language becomes much more sophisticated. In fact, solution to the problem may be described verbally to this "device" in ordinary English.

Thus in order to describe more fully what we mean by a method of solution, we must have some characterization of the device to be used. In general, we shall assume that the device operates by *discrete steps* which have a definite beginning and end. The operation within any one step may be continuous or not; it does not matter for our purposes. Also the steps may be performed sequentially or in parallel, and again this distinction is not important for the initial parts of our discussion. It is also necessary to assume that the device can communicate with the outside world through some *language*: it must be told what problem to solve and how to solve it, and must return the results to the user.

Let us then assume a device $M$ satisfying the above general requirements, and a problem $P$.

**Definition 1.1.** A *method of solution for problem P on device M* is a description in a language comprehensible to $M$ of discrete steps performable by $M$ and an ordering of these steps, such that given proper data, if $M$ performs the prescribed steps in the prescribed order, a solution to problem $P$ will result, if one exists. A method of solution will be called a *semi-algorithm* for $P$ on $M$ if the solution to $P$ (if one exists) appears after the performance of finitely many steps. A semi-algorithm will be called an *algorithm* if, in addition, whenever the problem has no solution the method enables the device to determine this after a finite number of steps and halt.

In other words, when using a semi-algorithm the device tries (rather blindly) to solve a problem and succeeds after finitely many steps whenever the problem is solvable. But when no solution exists, a semi-algorithm may cause the device being used to continue to search for a solution forever. An algorithm, on the other hand, also solves (either initially or concurrently) the related problem: does the given problem have a solution? It should also be pointed out that the "finite number of steps" may, in fact, be a very large finite number, and each step may be very lengthy and complicated, depending on the device used.

*Example 1.1.  Problem:*   Find $x$ such that $2 + x = 5$.
*Algorithm:*   Print "3."

*Example 1.2.  Problem:* Given $a$ and $b$ as data, find $x$ such that $a + x = b$.

We must now make some more specific assumptions about the device being used to solve the problem, in order that we may present the solution properly. We must not only assume that the device can read the data and print the answer, but also that it must be able to make comparisons and act according to the result of these comparisons. In addition, the device must have some method of storing data and intermediate results. Let us also assume that the device recognizes the non-negative integers, 0, 1, 2, 3, . . . . Now if the device can perform subtraction the solution to the problem is quite simple; but let us assume that the device can only perform addition. Thus we must form $a, a + 1, a + 2, a + 3, \ldots$, and test each of these against $b$. We might then have the following method of solution (to be written in a language which the device can read):

1. Read the data, $a$ and $b$.
2. Set the value of $x$ equal to 0.
3. Form $a + x$.
4. If $a + x = b$, print the value of $x$ and halt; otherwise continue.
5. Increase the value of $x$ by 1.
6. Return to step 3.

Notice that this is a semi-algorithm: if there is a non-negative $x$ such that $a + x = b$, this procedure will find it in a finite number of steps. However, if there is no such $x$ (for example, $4 + x = 3$), then the method of solution cycles forever.

We may change this into an algorithm by avoiding the cycling. If the machine can test for "greater than" this is very simple to do: between steps 3 and 4 add step 3.5.

3.5. If $a + x$ is greater than $b$, print "No answer." and halt; otherwise continue.

If the device can only test for equality, we can still develop an algorithm, provided that $a$ and $b$ are always non-negative integers (for this problem), but it is more complicated.

1. Read the data, $a$ and $b$.
2. Set the value of $x$ equal to 0.
3. Form $a + x$.
4. If $a + x = b$, print the value of $x$ and halt; otherwise continue.
5. Form $b + x$.
6. If $b + x = a$, print "No answer." and halt; otherwise continue.
7. Increase the value of $x$ by 1.
8. Return to step 3.

*Example 1.3.* There are, of course, many non-numerical problems which one can find algorithms to solve. Suppose, for example, that one must traverse a simple maze without loops. This is easily accomplished by choosing to follow a specific direction.

1. Whenever a branching of the maze is encountered, choose the right-most branch.
2. Whenever a dead-end is encountered, turn around, and continue.

By playing with a few such mazes, the reader can easily convince himself that a method of solution which is an algorithm for a particular device (such as a human being) can be based on these two rules. If the maze has loops the rules become more complex.

Mazes without loops are simple trees, in the sense of graph theory, and one finds these appearing frequently in searching problems, from looking up a phone number to tracking down an obscure technical reference. Of course, as people, we use more sophisticated search techniques; one of the major problems in instructing a mechanical device to solve a problem is to characterize these more sophisticated techniques in such a way that the device can utilize them.

EXERCISES

Find an algorithm to solve each of the following problems. Be sure to state your assumptions about the quantities involved and the capabilities of the device used.

1. Solve the quadratic equation $ax^2 + bx + c = 0$ for $x$, where $a$, $b$, and $c$ are integers.
2. Evaluate a $3 \times 3$ determinant whose entries are integers.
3. Traverse a maze with loops from a point $A$ to a point $B$.
4. For the first player to win or draw at the game of tic-tac-toe.

## 2. CHARACTERISTICS AND DESCRIPTIONS OF ALGORITHMS*

We have stipulated that an algorithm enables a device to solve a problem in a finite number of steps. We may classify algorithms according to the information which we have about this number of steps. First, there are some algorithms for which the number of steps is fixed, or has a fixed maximum. For example, if each step in an algorithm for playing tic-tac-toe is a play, there are at most nine steps. And in playing bridge, there are at most 319 bids in a hand, and exactly 52 cards are played. Any algorithm for playing bridge would be bound by these numbers. The algorithms in this class may or may not be cyclic in nature. If they are cyclic or iterative, there is then at least implicitly a counter to keep track of the number of steps carried out.

The other two classes of algorithms are generally iterative. For one of these classes the maximum number of steps involved is related to the data in such a way that it may be computed a priori. For example, the processing

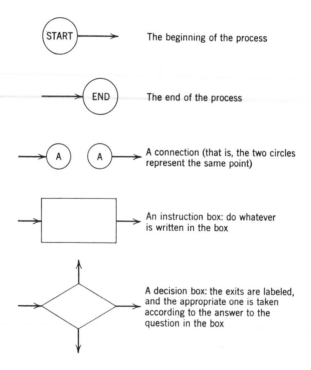

Figure 2.1. Basic flowchart notation.

* Much of what is written in this section applies to semi-algorithms as well.

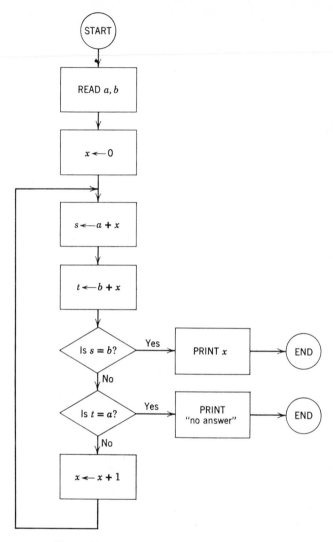

Figure 2.2. Flowchart for solving $a + x = b$.

of payroll checks involves at most $n \times k$ steps, where $n$ is the number of employees and $k$ is the maximum number of steps per check. In Example 1.2, the semi-algorithm involves exactly $4(b - a + 1)$ steps if there is a solution.

The maximum number of steps for the third class is also data-related, but in such a way that we cannot predict the number of steps involved.

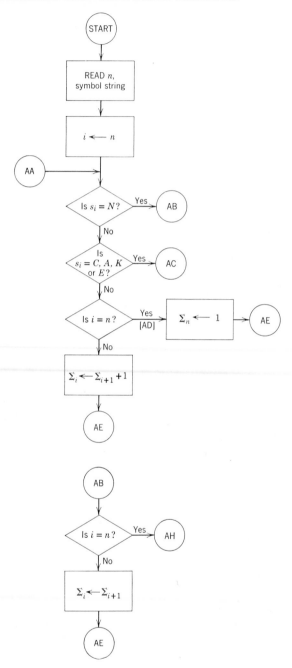

Figure 2.3. Flowchart for testing well-formed formulas in Polish notation.

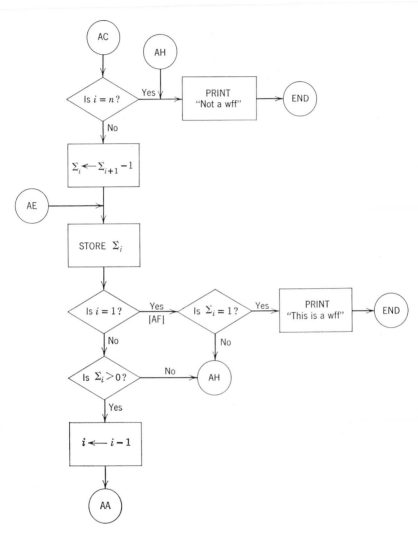

Figure 2.3—*Continued.*

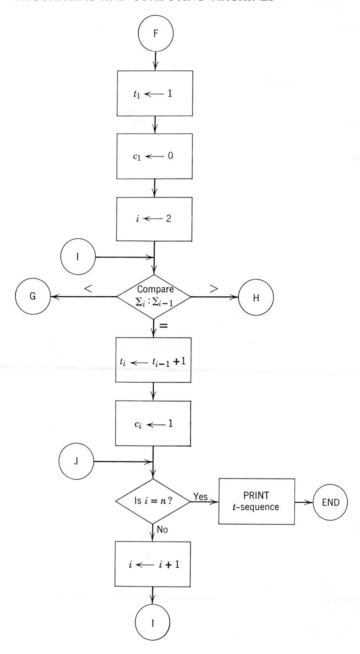

Figure 2.4.

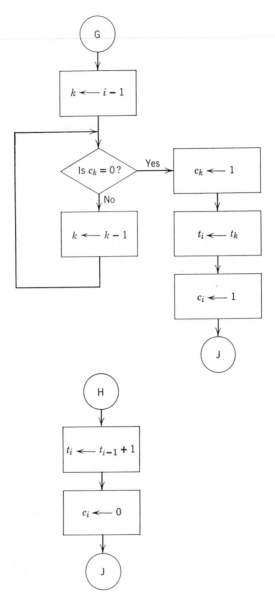

Figure 2.4—*Continued.*

Such algorithms are often related to searches in infinite sets, or to convergent processes. If, for example, we know that one integer has certain properties, then we can search all of the integers to find this one. But we may not be able to predict how long it will take to find this particular number. Similarly, in convergent processes we have a test which any acceptable result must satisfy, and we may know that the process will ultimately (in a finite number of steps) produce a result satisfying this test. Yet we may not be able to predict the number of steps necessary to produce this satisfactory result.

Observe that we can always transform a semi-algorithm into an algorithm of the first or second class by adding a counter. However, this generally transforms the problem from "Find a solution of . . ., if one exists," to "Find a solution of . . . in fewer than $n$ steps, if one exists." This change is an important one, for there exists a large class of problems for which there are semi-algorithms, but for which there are no algorithms except in the restricted sense of "solutions in fewer than $n$ steps."

What problems can be solved by algorithms? Intuitively speaking, the answer is any problem for which we can specify exactly a finite method of solution. Often the most convenient way to specify an algorithm is by means of a *flowchart* or *flow diagram*. This is a pictorial representation of the solution to a problem where the various steps to solve the problem are indicated by boxes with instructions written in them, and the order in which these instructions are to be executed is indicated by arrows. We illustrate this with a few examples using the notation shown in Figure 2.1.

*Example 2.1.  Problem:* Given $a$ and $b$ as data, find $x$ such that $a + x = b$. Figure 2.2 is the flowchart for the last algorithm given in Example 1.2 for the solution of this problem.

*Example 2.2.  Problem:* To determine whether or not a logical formula in Polish notation is well-formed. Let us call the symbol occurrences in the formula $s_1, \ldots, s_n$, numbering from the left, and assume that $n$ is given as part of the data. A flowchart for the algorithm discussed in Section 7 of Chapter 3 is given in Figure 2.3.

*Example 2.3.  Problem:* To develop the tree structure of a Polish notation formula. The algorithm which we gave for this in Chapter 3 depends on having the $\Sigma$ sequence developed in testing for a well-formed formula. Thus we will assume that we have the algorithm for that as it is shown in the flowchart of Example 2.2. Notice that we have explicitly included in the flowchart the instruction "STORE $\Sigma_i$." This insures that the $\Sigma$ sequence will be available for developing the tree structure. (Whether such a separate STORE instruction is needed depends on the device used to implement the algorithm.) We need to make one small change in the flowchart of the last problem. In Figure 2.3 we replace the (END) at the wff exit (but not the other one!) by a connector, say $\textcircled{F}$. The flowchart for the tree structure is then continued in Figure 2.4.

These examples illustrate the detail which must be specified for even relatively simple tasks. Although the principle used here may be applied to the solution of any task, one should be aware that for complex tasks—scheduling problems, information retrieval problems, marketing assignments—the flowchart and algorithm to solve the problem on a particular device will be very involved and still must be *completely* explicit. In particular, in tasks calling for judgment, whether intuitive or based on a complicated statistical analysis, the criteria for judgment must be reduced to definite "yes or no" criteria before an algorithm can be formulated to handle the job.

EXERCISES

Draw a flowchart for an algorithm solving each of the following problems.
1. Solve for $x$ and $y$ the two simultaneous equations $ax + by = c$ and $dx + ey = f$.
2. Determine all factors of an integer $n$.
3. Decide whose turn it is to hit next in a golf foursome.
4. Determine the alphabetic order of a set of $n$ words.

## 3. MARKOV ALGORITHMS

One might argue that for centuries men have solved problems just on the basis of intuitive judgments, and thus that we perhaps do not wish to use algorithms, but something rather different. Yet it develops that if we wish to effectively mechanize our problem and to use a computer-like device to aid in its solution we are forced to use algorithms. Numerous distinctly different approaches to problem solving have been proposed, and all of those which do not involve guesswork have been shown to be equivalent in the sense that they treat exactly the same set of problems (see Chapter 8). We wish in this and the next section to examine two of these approaches.

One of the more recent attacks on problem solving was the proposal by the Russian mathematician A. A. Markov in the early 1950's of what he called *normal algorithms*, which we shall call *Markov algorithms*. The general type of problem which Markov attacked is that of the transformation of strings of symbols: given string A, to transform it into string B in a mechanical way. This problem is an abstraction from many of our common problems. For example, the problem of adding two numbers $a$ and $b$ can be considered as the problem of transforming the string "$a + b$" into a string "$c$" representing the sum of $a$ and $b$. The problem of information retrieval can be thought of as the problem of transforming strings representing the request and the library of documents into strings representing the documents satisfying the request.

In transforming a string, we shall generally not want to operate on the entire string (which may be arbitrarily long) at once, but rather only on a small contiguous portion of it. Thus we shall assume that the device which is to utilize these algorithms is capable of recognizing the occurrences of a given substring within a given string. These occurrences may be several, and may in fact overlap. If we wish to distinguish a particular occurrence of a substring we may mark it with asterisks.

*Example 3.1.* The word *ratatattat* contains three occurrences of the string *tat*, namely *ra\*tat\*attat*, *rata\*tat\*tat*, and *ratatat\*tat\**. The first two of these occurrences overlap by one letter.

We shall consider these occurrences as numbered, and shall refer to the left-most occurrence of a string A in a string B as the *first occurrence of* A *in* B. We must call attention to one special string, namely the empty string, which contains no symbols. It plays a role analogous to that of the empty set. If a given string A contains $n$ symbols, the empty string $\Lambda$ is considered to have $n + 1$ occurrences in A: before the first symbol (the first occurrence of $\Lambda$), after the last symbol, and between every two adjacent symbols.

The transformations of which a Markov algorithm is composed are those that replace the first occurrence of a specified string A in the given string by another specified string B. A Markov algorithm then consists of a sequence of such transformations that are manipulated in a manner that we shall shortly make explicit. At this point it then appears that a Markov algorithm may well continue to apply these transformations forever, or may halt because none of the transformations may be applied to the string under consideration. We wish also to be able to specify explicitly that the algorithmic process halts at a certain point. Thus we have the two types of transformations defined in the following definition.

**Definition 3.1.** Let us consider strings of symbols from a given finite symbol set, called the *alphabet*. We suppose that the alphabet does not contain the symbols "$\rightarrow$" and ".". A *simple (Markov) production* is a string A $\rightarrow$ B, where A and B are strings in the alphabet. A *conclusive (Markov) production* is a string A $\rightarrow$ .B, where A and B are strings in the alphabet. In the production A $\rightarrow$ B (A $\rightarrow$ .B) the *antecedent* is A and the *consequent* B.

**Definition 3.2.** Let A $\rightarrow$ B (or A $\rightarrow$ .B) be a Markov production, where A and B are strings in an alphabet $\mathscr{A}$. Let S be a string of symbols in $\mathscr{A}$.

We say that the production is *applicable* to S if there is at least one occurrence of A in S. Otherwise the production is *not applicable* to S. If the first occurrence of A in S is P * A * Q, the *result* of applying the indicated production to S is the string PBQ.

*Example 3.2.* Let the alphabet be the English alphabet. The string S for this example is the string "*abactababrstc.*" Applying the production *act → bbb* to S we obtain the string *abbbbababrstc.* Applying the production *ba → .one* to S we obtain *aonectababrstc.* Applying the production *tab → Λ* to S we obtain *abacabrstc.* And applying *Λ → xz* to S we obtain *xzabactababrstc.* The production *abc → rst* is not applicable to S.

**Definition 3.3.** A *Markov algorithm* is a finite sequence $P_1, \ldots, P_n$ of Markov productions to be applied to strings in a given alphabet according to the following rules. Let S be a given string. The sequence is searched to find the first production $P_i$ whose antecedent occurs in S. If no such production exists, the operation of the algorithm halts without change in S. If there is a production in the algorithm whose antecedent occurs in S, the first such production is applied to S. If this is a conclusive production, the operation of the algorithm halts with no further change in S. If it is a simple production, a new search is conducted using the string S′ into which S has been transformed. If the operation of the algorithm ultimately ceases with a string S*, we say that S* is the *result* of applying the algorithm to S.

*Example 3.3.* Take the alphabet to be {*a, b, c, d*}. The algorithm we will consider is given below, with the productions numbered for reference.

1. *ad → .dc*
2. *ba → Λ*
3. *a → bc*
4. *bc → bba*
5. *Λ → a*

We apply this first to the string *dcb*, using a double arrow (⇒) to indicate the transformation effected.

$$dcb \Rightarrow adcb \quad (5)$$
$$\Rightarrow dccb \quad (1)$$

Since (1) is a conclusive production, the algorithm halts with the result *dccb*.

We next apply the algorithm to the string *dbc*.

$$dbc \Rightarrow dbba \quad (4)$$
$$\Rightarrow db \quad (2)$$
$$\Rightarrow adb \quad (5)$$
$$\Rightarrow dcb \quad (1)$$

Again, (1) is conclusive, so the result of applying the algorithm to *dbc* is the string *dcb*.

The third string we shall use is *bdc*.

$$
\begin{array}{lll}
bdc & \Rightarrow abdc & (5) \\
& \Rightarrow bcbdc & (3) \\
& \Rightarrow bbabdc & (4) \\
& \Rightarrow bbdc & (2) \\
& \Rightarrow abbdc & (5) \\
& \Rightarrow bcbbdc & (3) \\
& \Rightarrow bbabbdc & (4) \\
& \Rightarrow bbbdc & (2) \\
& \Rightarrow abbbdc & (5) \\
& \quad\quad \cdot \quad \cdot \quad \cdot
\end{array}
$$

The operation of the algorithm has not ceased at this point, and it is rather evident that the algorithm when applied to *bdc* will operate without ceasing, producing longer and longer strings of the form *b . . . bdc*.

The algorithm used in Example 3.3 has no purpose other than its use in the example. But if the concept of a Markov algorithm is to be useful we must show that we can accomplish meaningful tasks with these algorithms.

*Example 3.4.* Let the alphabet $\mathscr{A}$ be unspecified, and let A be a fixed string in this alphabet. We wish to transform an arbitrary string S into the string AS. This is easily accomplished with an algorithm consisting of one production, $\Lambda \rightarrow .A$.

Not all tasks are as easy to accomplish as that illustrated in Example 3.4. Suppose, for example, that we wished to transform S not into AS, but rather into SA. We cannot use the algorithm given in Example 3.4, for successive applications of this will produce the strings AS, AAS, AAAS, . . . (the first occurrence of $\Lambda$ is at the head of the word). Nor can we write the algorithm as $S \rightarrow .SA$, for there would need to be infinitely many productions with no first production! In fact, because the productions are always applied to the *first* occurrence of A in B, there is difficulty any time we wish to operate with the second, or third, or last occurrence. We overcome this difficulty by the use of special marker symbols which are not part of the given alphabet. Thus we will speak of an algorithm *over* an alphabet, meaning that its productions contain symbols from the alphabet and certain specified other symbols, such as markers. By use of these markers, we may mark any particular point within a string and then operate at that point.

*Example 3.5.* Let the given alphabet be $\mathscr{A}$, and let $\alpha$ be a marker (not an element of $\mathscr{A}$). If S is a string in $\mathscr{A}$, the result of applying the following algorithm to S is the string SA.

$$\alpha\zeta \to \zeta\alpha \quad (\zeta \in \mathscr{A})$$
$$\alpha \to .A$$
$$\Lambda \to \alpha$$

Since S initially does not contain $\alpha$, the third production is used to obtain $\alpha$S. The first production is then used to move the $\alpha$ past the symbols in S. If S contains $n$ occurrences of symbols, then after $n$ steps we obtain the string S$\alpha$. At this point the first production no longer applies, and the second produces SA. Since this production is conclusive, the string SA is then the result.

In Example 3.5 we have introduced a new notation. Namely, in the first production we have used the variable $\zeta$ which ranges over the symbols of $\mathscr{A}$. Thus the first line is not really a production, but rather a *production schema*, denoting all the productions which can be obtained by substituting symbols of $\mathscr{A}$ for $\zeta$. (Compare the discussion of axiom schemata in Chapter 3, Section 9.)

Because of the manner in which the Markov algorithms are used, the order in which the productions are written is vital. For example, if the first two lines of the algorithm schema in Example 3.5 were interchanged, the result would be to transform S into AS, rather than SA, and the productions represented by $\alpha\zeta \to \zeta\alpha$ would never be used. A little thought should convince the reader that the production schemata each represent sections of the algorithm in which the order of the individual productions is not critical since at most one of the represented productions applies and they all do the same thing, in different contexts.

We conclude this section with several examples of tasks which can be accomplished by Markov algorithms. The development of this subject goes far beyond our introduction here, and the interested reader is referred to Markov's book [13] for further details. In particular, any task which can be accomplished by the use of algorithms in the more liberal sense of Section 1 can be accomplished by the use of Markov algorithms.

In the following examples the alphabet $\mathscr{A}$ will be left unspecified, except that it contains none of the markers or special symbols which are explicitly stated.

*Example 3.6.* This algorithm transforms every string into the empty string.

$$\zeta \to \Lambda \quad (\zeta \in \mathscr{A})$$

In operation, this algorithm successively picks off the first letters of a string, as long as any letters remain. When the string becomes the empty string the process halts, since there is no transformation whose antecedent is the empty string.

*Example 3.7.* This algorithm leaves the empty string unchanged, but deletes the first letter of any nonempty string and then halts. Marker: $\alpha$.

$$\alpha\zeta \rightarrow .\Lambda \qquad (\zeta \in \mathscr{A})$$
$$\alpha \rightarrow .\Lambda$$
$$\Lambda \rightarrow \alpha$$

*Example 3.8.* Often it is useful to know the number of symbols in a string. This is easily accomplished in tally notation by replacing every symbol by a tally mark. Special symbol: 1.

$$\zeta \rightarrow 1 \qquad (\zeta \in \mathscr{A})$$

Since "1" is not an element of $\mathscr{A}$ the operation ceases when every symbol has been transformed.

*Example 3.9.* At times one wishes to discard a portion of a symbol string, as one would discard data after computing the answer. The following algorithm discards everything to the left of the special symbol $\alpha$.

$$\zeta\alpha \rightarrow \alpha \qquad (\zeta \in \mathscr{A})$$
$$\alpha \rightarrow .\Lambda$$

*Example 3.10.* In almost every problem there is some point at which a decision must be made, dependent on the results of the calculation up to that point. We now present a Markov algorithm for making such a decision. An arbitrary string in the given alphabet is examined to determine whether it is a specified string A. If it is, the entire string is replaced by the string B; otherwise the entire string is replaced by the string C. Marker: $\alpha$.

$$\zeta\alpha \rightarrow \alpha\zeta \qquad (\zeta \in \mathscr{A})$$
$$\alpha\zeta \rightarrow \alpha \qquad (\zeta \in \mathscr{A})$$
$$\alpha \rightarrow .C$$
$$A\zeta \rightarrow \alpha \qquad (\zeta \in \mathscr{A})$$
$$\zeta A \rightarrow \alpha \qquad (\zeta \in \mathscr{A})$$
$$A \rightarrow .B$$
$$\Lambda \rightarrow \alpha$$

If the given string, P, does not contain an occurrence of the string A, the last production introduces an $\alpha$, and then the second and third production schemata erase P and replace it by C. If P contains an occurrence of A but is *not* A, either the fourth or the fifth production schema is used to introduce the $\alpha$; the first schema moves $\alpha$ to the left end of P and then the second and third operate as before. Finally, if the string P is actually A, the sixth production applies, and P is transformed into B.

Notice that the productions in the algorithm of Example 3.10 refer directly to the string A, which might be quite long. Since A is known a priori this is permissible: we could always replace such a reference by a letter by letter search for A.

*Example 3.11.* Another procedure which is quite common is that of doubling or duplicating a string. Often we wish to perform transformations which destroy a string, but which are only tentative in nature: at some point we may decide that the transformations are wrong and wish to begin anew. Thus we must be able to save a copy of the original string to which we can return. Given a string P, the following algorithm produces the string PP before it halts. Markers: $\alpha$, $\beta$, $\gamma$.

$$\zeta\eta\beta \to \eta\beta\zeta \qquad (\zeta, \eta \in \mathscr{A})$$
$$\alpha\zeta \to \zeta\beta\zeta\alpha \qquad (\zeta \in \mathscr{A})$$
$$\beta \to \gamma$$
$$\gamma \to \Lambda$$
$$\alpha \to .\Lambda$$
$$\Lambda \to \alpha$$

The reader should try applying this algorithm to a few strings in order to understand its operation.

*Example 3.12.* As our final example, we give a projection algorithm. We are concerned with strings of the form $A_1\alpha A_2\alpha \ldots \alpha A_n$, where $\alpha$ is a special symbol. The task is to isolate from this string the substring $A_i$, where $i$ and $n$ are known. This is accomplished by erasing $A_1, A_2, \ldots, A_{i-1}$, passing over $A_i$, and then erasing the remaining portion of the string. Markers: $\beta_1, \ldots, \beta_i, \gamma$.

$$\beta_1\zeta \to \beta_1 \qquad (\zeta \in \mathscr{A})$$
$$\beta_1\alpha \to \beta_2$$
$$\beta_2\zeta \to \beta_2 \qquad (\zeta \in \mathscr{A})$$
$$\beta_2\alpha \to \beta_3$$
$$\vdots$$
$$\beta_{i-1}\zeta \to \beta_{i-1} \qquad (\zeta \in \mathscr{A})$$
$$\beta_{i-1}\alpha \to \beta_i$$
$$\beta_i\zeta \to \zeta\beta_i \qquad (\zeta \in \mathscr{A})$$
$$\beta_i\alpha \to \gamma$$
$$\gamma\zeta \to \gamma \qquad (\zeta \in \mathscr{A})$$
$$\gamma\alpha \to \gamma$$
$$\gamma \to .\Lambda$$
$$\Lambda \to \beta_1$$

## EXERCISES

Develop Markov algorithms for the following tasks.

1. An algorithm which transforms any string into itself.

2. An algorithm which is applicable to no string. That is, no matter what string is used, the operation of the algorithm never halts.
3. Let $a \in \mathscr{A}$ be specified. It is desired to transform any string by deleting all occurrences of $a$.
4. Let A be a specified string of symbols in $\mathscr{A}$. It is desired to replace any string P by the string A.
5. Let $\alpha$ be a special symbol (not in $\mathscr{A}$). Transform P$\alpha$Q into P.
6. Let $\alpha$ and $\beta$ be special symbols. Transform P$\alpha$Q$\beta$R into Q.
7. Let S and A be such that the first occurrence of A in S is P$*$A$*$Q. Transform S (that is, PAQ) into P$*$A$*$Q.
8. Let $\alpha$ and $\beta$ be special symbols. Transform P$\alpha$Q into P$\alpha$Q$\beta$P.
9. Let $x$ and $y$ be tally representations of positive integers. Transform the string $x+y$ into a string representing the sum of $x$ and $y$.
10. Transform $x*y$ into a string representing the product of $x$ and $y$.
11. Transform the tally representation of a positive integer into the corresponding decimal numeral.
12. Determine what the following algorithm does, and explain why it does not double a string (see Example 3.11). Markers: $\alpha$, $\beta$.

$$\zeta\eta\beta \to \eta\beta\zeta \qquad (\zeta, \eta \in \mathscr{A})$$
$$\alpha\zeta \to \zeta\beta\zeta\alpha \qquad (\zeta \in \mathscr{A})$$
$$\beta \to \Lambda$$
$$\alpha \to .\Lambda$$
$$\Lambda \to \alpha$$

## 4. TURING MACHINES

We turn now to a more classic approach to the question of solving problems. In 1936 the British mathematician, Alan Turing, wrote a paper in which he developed a mechanical method for solving problems, which has since come to be known as a *Turing machine* [19]. Several different but equivalent models of Turing machines have been proposed. We shall discuss only one of these. After giving a formal definition, we shall give an intuitive description of the operation of this type of Turing machine and several examples of its use.

**Definition 4.1.** Let $\mathscr{Q} = \{q_0, q_1, \ldots, q_n\}$, $\mathscr{S} = \{s_0, s_1, \ldots, s_m\}$, and $\mathscr{M} = \{R, L\}$ be finite sets known respectively as the sets of *states*, *symbols*, and *moves*. An *expression* is a finite sequence of signs all of which except one are symbols (that is, elements of $\mathscr{S}$), the one exception being a state (that is, an element of $\mathscr{Q}$). A *quintuple* is a sequence of five elements $(q_i, s_j, s_k, m, q_l)$ where $q_i$ and $q_l$ are states, $s_j$ and $s_k$ are symbols, and $m$ is a move. A *Turing machine* is a finite set of quintuples, no two of which have the same initial pair of elements $q_i$ and $s_j$.

A Turing machine transforms expressions into other expressions according to the following rules. Let P and Q represent arbitrary strings of symbols (in $\mathscr{S}$), possibly empty, and let $s' \in \mathscr{S}$. We consider $s_0$ to be a special symbol, called a *blank*, and usually denote it either by $\square$ or by 0.

*Case 1.* Quintuples of the form $q_i\, s_j\, s_k\, \text{R}\, q_l$.

Expressions of the form $Pq_is_js'Q$ are transformed into expressions $Ps_kq_is'Q$.

Expressions of the form $Pq_is_j$ are transformed into expressions $Ps_kq_l\square$. Other expressions are not transformed.

*Case 2.* Quintuples of the form $q_i\, s_j\, s_k\, L\, q_l$.

Expressions $Ps'q_is_jQ$ are transformed into expressions $Pq_is's_kQ$.

Expressions $q_is_jQ$ are transformed into expressions $q_l\square s_kQ$. Other expressions are not transformed.

A Turing machine when applied to an expression is allowed to operate until it reaches a state-symbol pair $q_is_j$ for which no quintuple is defined. The expression at that time may be considered the *result* of applying the Turing machine to the initial expression, although more commonly the result is considered to be the sequence of symbols at that time, that is, ignoring the final state.

*Example 4.1.*

| | | | | | | | |
|---|---|---|---|---|---|---|---|
| $q_0$ | $\square$ | $\square$ | $L\ q_4$ | $q_1\ c$ | $c$ | $R\ q_0$ | $q_4\ d\quad d\ L\ q_4$ |
| $q_0\ a$ | $a$ | $R\ q_1$ | | $q_1\ d$ | $c$ | $L\ q_2$ | $q_5\ \square\ b\ R\ q_6$ |
| $q_0\ b$ | $b$ | $R\ q_0$ | | $q_2\ a$ | $d$ | $L\ q_7$ | $q_5\ a\quad a\ L\ q_4$ |
| $q_0\ c$ | $c$ | $R\ q_0$ | | $q_3\ a$ | $a$ | $L\ q_5$ | $q_5\ b\quad b\ L\ q_4$ |
| $q_0\ d$ | $d$ | $R\ q_0$ | | $q_4\ \square$ | $\square$ | $R\ q_7$ | $q_5\ c\quad c\ L\ q_4$ |
| $q_1\ \square$ | $\square$ | $L\ q_3$ | | $q_4\ a$ | $a$ | $L\ q_4$ | $q_5\ d\quad d\ L\ q_4$ |
| $q_1\ a$ | $a$ | $R\ q_1$ | | $q_4\ b$ | $b$ | $L\ q_4$ | $q_6\ a\quad c\ L\ q_6$ |
| $q_1\ b$ | $b$ | $R\ q_0$ | | $q_4\ c$ | $c$ | $L\ q_4$ | $q_6\ b\quad b\ L\ q_7$ |

This Turing machine operates on the symbol set $\{\square\ a, b, c, d\}$ (or any set containing this as a subset). There are references among these quintuples to a state $q_7$, yet none of the quintuples begins in $q_7$. Thus $q_7$ is, in effect, a "stop" state. Notice also that certain other state-symbol pairs, for example, $q_2b$, do not begin any of the quintuples listed. Thus although $q_2$ is not always a stop state, the machine will halt if while in state $q_2$ it encounters the symbol $b$.

To illustrate the operation of this machine, consider the expression $q_0bcadc$. The machine will transform this successively into $bq_0cadc$, $bcq_0adc$, $bcaq_1dc$, $bcq_2acc$, and $bq_7cdcc$, at which point it will halt. Thus the machine has transformed the symbol string $bcadc$ into the string $bcdcc$. This corresponds to the Markov production $ad \rightarrow .dc$. The reader is invited to apply the Turing machine to other expressions of his own choosing.

A convention which is useful in working with Turing machines is that of a tabular representation. In this form, the quintuple $q_i\,s_j\,s_k\,m\,q_l$ is represented by the table entry $s_k\,m\,q_l$ with coordinates $q_i$ and $s_j$. In this form the Turing machine of Example 4.1 is shown in Table 4.1.

### Table 4.1

The Turing machine of Example 4.1

|       | □        | $a$     | $b$     | $c$     | $d$     |
|-------|----------|---------|---------|---------|---------|
| $q_0$ | □ $Lq_4$ | $aRq_1$ | $bRq_0$ | $cRq_0$ | $dRq_0$ |
| $q_1$ | □ $Lq_3$ | $aRq_1$ | $bRq_0$ | $cRq_0$ | $cLq_2$ |
| $q_2$ |          | $dLq_7$ |         |         |         |
| $q_3$ |          | $aLq_5$ |         |         |         |
| $q_4$ | □ $Rq_7$ | $aLq_4$ | $bLq_4$ | $cLq_4$ | $dLq_4$ |
| $q_5$ | $bRq_6$  | $aLq_4$ | $bLq_4$ | $cLq_4$ | $dLq_4$ |
| $q_6$ |          | $cLq_6$ | $bLq_7$ |         |         |

Let us now look at an intuitive interpretation of a Turing machine. Let us consider a machine consisting of a black box with a read-write head designed to scan a tape which is divided into squares. The tape may be considered either infinite, or arbitrarily extendable in both directions (see Figure 4.1). Within the black box the device may take on one of a number of internal configurations or states. These provide the machine with a limited amount of memory. In operation, the read-write head scans a square on the tape. Depending on its present internal state and the symbol which it finds on the square, it prints another symbol on that square, moves one square to the right or left, and changes its state. The formal assumption that an expression is finite in length corresponds to the informal assumption that at any one time the tape is blank except for a finite number of symbols on it.

*Example 4.2. A counter.* We suppose that on the tape there is a sequence of ones and zeros bounded on the left and on the right by the symbol " $ ".

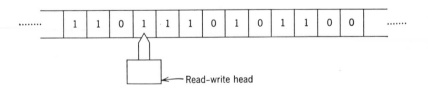

Figure 4.1. The basic Turing machine.

**Table 4.2**

A counter

| | $\$$ | 0 | 1 | 2 | 3 | 4 | 5 | 6 | 7 | 8 | 9 | $a$ | $\square$ |
|---|---|---|---|---|---|---|---|---|---|---|---|---|---|
| $q_0$ | $\$Rq_1$ | $0Rq_1$ | $aLq_2$ | | | | | | | | | | |
| $q_1$ | $\$Rq_6$ | $0Lq_2$ | $1Lq_2$ | | | | | | | | | | |
| $q_2$ | $\$Lq_3$ | $1Rq_4$ | $2Rq_4$ | $3Rq_4$ | $4Rq_4$ | $5Rq_4$ | $6Rq_4$ | $7Rq_4$ | $8Rq_4$ | $9Rq_4$ | $0Lq_3$ | | |
| $q_3$ | | | | | | | | | | | | | $1Rq_4$ |
| $q_4$ | $\$Rq_5$ | $0Rq_4$ | | | | | | | | | | | |
| $q_5$ | | $0Rq_5$ | $1Rq_5$ | | | | | | | | | $1Rq_1$ | |

It is desired to count the number of ones in this sequence and to print to the left of the left-most " $ " the decimal numeral representing this number of ones. We wish to leave the original string unchanged, and assume that we have a symbol "$a$" which does not occur, and hence is available to use as a marker. If we assume that initially the Turing machine is in state $q_0$ and is scanning the left-most " $ ", then the machine shown in Table 4.2 performs the desired task. It accomplishes this by searching to the right to find a "1" which it replaces by "$a$". It then returns left and adds one to the count which it is building up. After this, the machine locates the "$a$", replaces it with a "1", and resumes the search. Location of the right-most " $ " halts the process.

Table 4.3
An adder

| | 0 | 1 |
|---|---|---|
| $q_0$ | $0Rq_7$ | $0Rq_1$ |
| $q_1$ | $0Rq_2$ | $1Rq_1$ |
| $q_2$ | $0Rq_3$ | $1Rq_2$ |
| $q_3$ | $1Lq_4$ | $1Rq_3$ |
| $q_4$ | $0Lq_5$ | $1Lq_4$ |
| $q_5$ | $0Lq_6$ | $1Lq_5$ |
| $q_6$ | $1Rq_0$ | $1Lq_6$ |
| $q_7$ | $0Rq_{12}$ | $0Rq_8$ |
| $q_8$ | $0Rq_9$ | $1Rq_8$ |
| $q_9$ | $1Lq_{10}$ | $1Rq_9$ |
| $q_{10}$ | $0Lq_{11}$ | $1Lq_{10}$ |
| $q_{11}$ | $1Rq_7$ | $1Lq_{11}$ |

*Example 4.3. An adder.* Our symbol set now consists of just two symbols, a zero and a one. We wish to transform the string

$$\underbrace{1\ldots1}_{m}0\underbrace{1\ldots1}_{n} \quad \text{into the string} \quad \underbrace{1\ldots1}_{m}0\underbrace{1\ldots1}_{n}0\underbrace{1\ldots1}_{m+n}$$

That is, considering the original string as a tally representation of a pair of positive integers we wish to form the sum of these numbers while retaining the original numbers. It is permissible to use markers as was done in the last example, but not necessary. In fact, the machine shown in Table 4.3 performs this task without the benefit of markers. We assume that the machine is initially in state $q_0$ and scanning the leftmost "1".

If the reader tries to write an adder for solving the problem of Example 4.3 using markers, he will discover that it can be done using a machine with fewer states. In general there is a trade-off between states and symbols: if a marker is not used to locate a specific point, then additional states must be used to keep track of the location of the read-write head relative to that point. However, we cannot reduce the number of symbols below two.

*Example 4.4. A number recognizer.* This is a fragment of a Turing machine, designed to terminate its operation in one of a hundred different states, depending on which of the numbers $00, 01, \ldots, 99$ is printed on the tape. The machine, shown in Table 4.4, is initially in state $q_0$ and scanning the left-most digit of the pair.

As is evident from this last example, we can describe a Turing machine which will distinguish between an arbitrary finite number of numerals or other symbol strings, and act accordingly. Thus once we settle on a symbol set and an exact method of writing descriptions of problems, we can build a Turing machine which will distinguish between problems and react accordingly, that is, try to solve the problem described.

**Table 4.4**

A number recognizer

|       | 0          | 1          | 2          | ... | 9          |
|-------|------------|------------|------------|-----|------------|
| $q_0$ | $0Rq_0$    | $1Rq_1$    | $2Rq_2$    |     | $9Rq_9$    |
| $q_1$ | $0Rq_{10}$ | $1Rq_{11}$ | $2Rq_{12}$ |     | $9Rq_{19}$ |
| $q_2$ | $0Rq_{20}$ | $1Rq_{21}$ | $2Rq_{22}$ |     | $9Rq_{29}$ |
| .     | .          | .          | .          | .   | .          |
| $q_9$ | $0Rq_{90}$ | $1Rq_{91}$ | $2Rq_{92}$ |     | $9Rq_{99}$ |

In particular, we have an exact method for describing Turing machines and the data upon which they operate. Thus one can construct a Turing machine M which will accept as input the description of an arbitrary Turing machine T and some data, and will perform the same calculation on the data that T would. Such a machine M is known as a *universal Turing machine.*

This does not mean that we can solve all problems with a universal Turing machine. In particular, feeding the description of M itself into M leads to unsolvable problems. A more easily visualized unsolvable problem is the "halting problem," which is discussed in the next section.

In addition to other models of the Turing machines, a number of quite different "machines" have been proposed and studied by various specialists. These devices generally are known as *automata* of various types, and are often more limited in their capabilities than is a Turing machine.

Nevertheless the study of these devices yields important insights into the relative complexity of different kinds of problems.

EXERCISES

1. Suppose that a tape for a Turing machine has on it exactly one finite block of symbols, that block being bounded on the left and on the right by the symbol "$". Design a Turing machine which can begin at any location on the tape, search for this block of symbols, and halt at the left-most $. You may assume that the symbols "*a*", "*b*, and "*c*" are not on the tape and may be used in any way you wish.
2. Describe a Turing machine to distinguish between odd and even numbers written in (a) tally notation (111111111) and (b) decimal notation. The machine is to print on the tape to the right of the number the symbol "A" if the number is even, and "B" if it is odd.
3. Describe a Turing machine which will find the remainder upon dividing a number by three, when the notation is (a) tally notation and (b) decimal notation.

   (*Note:* There are easy and hard ways to solve this problem.)
4. Suppose that a tape has on it two words in the alphabet {*a, b, c, d*}. The words are separated by one blank, and the remainder of the tape is blank. Design a Turing machine which will check the words and if necessary rearrange them so that the left-most word precedes the right-most alphabetically.

## 5. THE BUSY BEAVER AND HALTING PROBLEMS

Turing machines are constructed to perform specific tasks such as addition or multiplication. Part of the construction is the tacit assumption of a standard format for the input string, such as we specified in Examples 4.2 and 4.3. Thus one is naturally led to question the performance of the machine on a non-standard input string. What, for example, would happen if the counter we designed encountered symbols other than 0, 1, and *a* between the dollar signs; or what would happen if the right-most dollar sign was missing? This is the *halting problem*: given a Turing machine and an arbitrary tape, to determine whether or not the machine would eventually halt using the given tape as input. This and the related Busy Beaver problem have been shown to be unsolvable by any Turing machine (or algorithm). That is, it is not possible to design an algorithm which will solve this problem. The essential word here is "eventually." It is easy to determine whether or not a given machine using a given tape will halt within 1,479,641 or any other given number of steps: just try to run the machine for 1,479,642 steps. But with "eventually," we have no limit on the possible number of steps which may occur.

There are only a finite number of Turing machines of a given size (that is, number of states and symbols). For example, if we allow $n$ states (not counting the halt state), two moves, and two symbols (0 and 1), then each block in the table describing a machine may be filled in $4(n + 1)$ ways (the extra one is for the halt state). Since there are $2n$ blocks in the table, if we require that each block be filled there are exactly $N = (4(n + 1))^{2n}$ $n$-state two-symbol Turing machines. The *Busy Beaver problem* (of class $(n, 2)$) is to determine which of these machines will, when started with a blank tape, halt with the highest possible number of 1's on its tape. This is thus a specialized halting problem, which has been shown by Rado [16] to be unsolvable. Nevertheless, some work has been done on this with interesting results [9]. It is known that for two-symbol machines, the highest possible number of 1's obtainable with a halting machine of 1 state is 1, 2 states—4, and 3 states—6. Table 5.1 shows one of the 3-state machines which will halt with six 1's. Four other such machines exist.

### Table 5.1

A machine solving the three-state Busy Beaver problem

|       | 0        | 1        |
|-------|----------|----------|
| $q_0$ | $1Rq_1$  | $1Rq_0$  |
| $q_1$ | $1Lq_2$  | $1Rq_3$  |
| $q_2$ | $1Rq_0$  | $1Lq_1$  |

For Turing machines having more than three states or operating on more than two symbols, the maximum possible score is not known. Nor has anyone solved the related problem of determining the maximum number of moves or shifts which is possible in a machine which halts. The known results are given in Table 5.2, where $\Sigma(n)$ denotes the maximum possible score, and $SH(n)$ denotes the maximum possible number of shifts.

To indicate the magnitudes which must be considered in this problem, let us look at the 100-state machines. There are $163,216^{100}$ of these, some of which will halt when started with a blank tape, and some of which will not. It is known that one of these will halt with $(((7!)!)!)!$ or approximately $10^{10^{10^{15000}}}$ 1's on the tape. Thus the maximum number of ones attainable is at least that large, and probably considerably larger. Yet if we use ten billion years as an estimate of the age of the universe and assume that one billion 1's can be printed per second (somewhat faster than current digital computers), only approximately $3.15 \times 10^{26}$ of these 1's could have been printed since the universe began.

**Table 5.2**

The known results in the Busy Beaver problem*

| $n =$ | 1 | 2 | 3 | 4 | 5 | 6 | 7 | 8 |
|-------|---|---|---|---|---|---|---|---|
| Two-symbol machines | | | | | | | | |
| $\Sigma(n)$ | $= 1$ | $= 4$ | $= 6$ | $\geq 13$ | $\geq 17$ | $\geq 35$ | $\geq 22{,}961$ | $\geq 3(7.3^{92} - 1)/2$ |
| $SH(n)$ | | | $= 21$ | $\geq 107$ | | | | |
| Three-symbol machines | | | | | | | | |
| $\Sigma(n)$ | | | $\geq 12$ | | | | | |
| $SH(n)$ | | | $\geq 57$ | | | | | |

* These results were communicated to the author in February 1966 by C. Y. Lee of Bell Telephone Laboratories, and are due to Lee, Tibor Rado, Shen Lin, Patrick Fischer, Milton Green, and David Jefferson.

### EXERCISES

1. Solve the Busy Beaver problem for 1-state two-symbol machines. That is, find a Turing machine with just one state other than the stop state which will print one 1 and halt.
2. Solve the Busy Beaver problem for 2-state two-symbol machines.
3. Find a solution to the 3-state two-symbol Busy Beaver problem other than that given in the text.

## 6. DIGITAL COMPUTERS

Both Markov algorithms and Turing machines are very satisfying theoretically. They involve easily understood principles, and it is quite simple to show that they are equivalent to each other and what their relationship is to the other computational systems which have been proposed. Yet their very simplicity makes them unsuited for most practical computational needs: witness the labor involved in multiplying two positive integers by either of these devices. We wish now to examine a device which is close to modern digital computers, and to see how we would solve problems using it. We caution the reader that many considerations such as the limitations of register sizes, timing problems, and access to input-output devices are omitted, although they are vital to the accomplishment of computing. Our purpose is to examine the logic behind the method of solution rather than the details used to make the machine behave.

The "computer" which we shall discuss is an extremely simplified abstraction of current digital computers. Actual computers often have over 100 operation codes, several index registers to aid in addressing, different types of storage, and so forth. We will assume a single storage unit, which is symbolically addressable. That is, we may refer to a quantity X in storage as "X": we do not need to know its exact location in storage. Two special registers, the *accumulator* and the *MQ* (multiplier-quotient) registers, will be used for all operations. These are not considered part of storage.

We shall assume twelve instructions, which are single-address and fall into three categories. By "single-address" we mean that an instruction refers to only one storage location. Thus we cannot in a single instruction add two numbers, since we cannot refer to both within the same instruction.

The first four instructions deal with the transfer of information between storage and the two special registers. CLA ("clear and add") and LDQ ("load the MQ") are used to move information from storage to the accumulator and MQ respectively. They are considered to be non-destructive. That is, if a quantity X is located in storage, then execution of the instruction CLA X produces a copy of X in the accumulator, while retaining X in storage. The instructions STO ("store the accumulator") and STQ ("store the MQ") enable us to move information from the accumulator and MQ respectively into storage. Again, these are non-destructive: if Y is located in the MQ, then STQ Y produces a copy of Y in storage in addition to the copy of Y left in the MQ.

There are four arithmetic instructions: ADD, SUB, MPY, DIV. Since these are single-address instructions and the arithmetic operations involve two operands, it is necessary to have one of the operands previously located in the accumulator, or for multiplication in the MQ. This is not a great restraint on programming since many times these operations are used sequentially, and one of the operands is already in the accumulator or the MQ from the previous operation. The basic sequences for producing sums, differences, products, and quotients, and placing them in storage are as follows.

$X + Y = Z$: CLA X, ADD Y, STO Z. (Only the accumulator is used.)

$X - Y = Z$: CLA X, SUB Y, STO Z.

$X \times Y = Z$: LDQ X, MPY Y, STQ Z (or STO Z). (The multiplier is held in the MQ register. The product expands to twice the length of a normal storage register—for example, $76 \times 24 = 1824$—and is held both in the MQ and the accumulator. The user must know whether to save the high order part, which is in the MQ, or the low order, which is in the accumulator.)

$X \div Y = Z$: CLA X, DIV Y, STQ Z. (In division, the quotient is left in the MQ with the remainder in the accumulator. Thus the remainder is also available if it is needed.)

Again the reader is cautioned that in programming for an actual computer, the arithmetic processes are complicated by the fact that the programmer must be aware of the possibilities of overflow, division by zero, and so forth. However, the basic instructions are as we have given.

Finally there are a group of four control instructions. In a digital computer the instructions are located in storage along with the data and in the same form. Thus they can be manipulated and referred to like the data. In particular, although normally a control circuit automatically sequences the instructions, it is possible to name an instruction and force it to be used next. This is the purpose of the control instructions. The first of these is an unconditional transfer, TRA ("transfer to . . ."). The execution of the instruction TRA X does nothing arithmetically, but causes the next instruction executed to be the one labeled "X". The sequencing then proceeds automatically from that point without reverting to the former sequencing.

There are two conditional transfer instructions, TRZ ("Transfer on zero") and TRP ("transfer on plus"), which cause transfers of control whenever the contents of the accumulator are zero or positive, respectively.

Finally, there is the instruction HLT ("halt"), which causes the computer to cease operation. It requires no operand.

We shall assume that the accumulator and MQ registers are addressable by the names "ACC" and "MQ". Thus, for example, LDQ ACC would cause the contents of the accumulator to be placed in the MQ. However, it is not permitted to transfer control to either the accumulator or the MQ.

With this brief introduction, we shall turn to three examples of the manner in which one would realize algorithms on a digital computer. It should be remembered that an actual computer has a much larger repertoire of instructions, and thus is able to handle these computations in a more sophisticated manner.

*Example 6.1.* The problem is to solve the system of two linear equations (6.1) for $x$ and $y$. We assume that the coefficients are stored as A, B, C, D, E, and F, and that the results will be called X and Y in the computer. In actual practice we would need to read the coefficients into storage, and print out the results, but we ignore these steps here.

$$ax + by = c$$
$$dx + ey = f \qquad (6.1)$$

The solution of these equations is given by equations (6.2).

$$x = \frac{ce - bf}{ae - bd}$$

$$(6.2)$$

$$y = \frac{af - cd}{ae - bd}$$

In the program given below we have assumed that the coefficients are such that the significant parts of the products occur in the MQ. The comments are explanation for the reader, and not part of the program.

```
LDQ  A
MPY  E
STQ  TEMP     Store the product ae and name it TEMP.
LDQ  B
MPY  D
CLA  TEMP     At this point bd is in the MQ and ae in the accumulator.
SUB  MQ
STO  DENOM    The denominator of the fractions has been computed.
LDQ  C        This begins the computation for x.
MPY  E
STQ  TEMP     TEMP now contains the product ce.
LDQ  B
MPY  F
CLA  TEMP
SUB  MQ       This computes the numerator for x.
DIV  DENOM
STQ  X
LDQ  A        This begins the computation for y.
MPY  F
STQ  TEMP
LDQ  C
MPY  D
CLA  TEMP
SUB  MQ       The numerator for y.
DIV  DENOM
STQ  Y
HLT
```

*Example 6.2.* The algorithm or program in the previous example consists of three parts which are nearly the same except for the data used.

Whenever the computation in various parts of a program is the same, but applied to different data, it is convenient to be able to cycle back through the computation rather than write out and provide storage space for several copies. In this example we program the algorithm for determining whether or not a Polish formula is well-formed, using the ability of the computer to cycle through a computation several times. The flowchart for this algorithm was given in Figure 2.3. We assume that we have in storage the data NUM, S(1), S(2), ..., S(NUM), and the constants C, A, K, E, and N. Notice that the data in this case are alphabetic. Nevertheless, in the computer they are represented in the same form as numerical data (and instructions), and hence we can perform numerical operations on them. We also need the numerical constant 1 in storage.

|    |     |        |                                                        |
|----|-----|--------|--------------------------------------------------------|
|    | CLA | NUM    |                                                        |
|    | STO | I      |                                                        |
| AA | CLA | S(I)   |                                                        |
|    | SUB | N      |                                                        |
|    | TRZ | AB     | Transfer to instruction AB if S(I) = N.                |
|    | ADD | N      |                                                        |
|    | SUB | C      |                                                        |
|    | TRZ | AC     | Transfer to instruction AC if S(I) = C.                |
|    | ADD | C      |                                                        |
|    | SUB | A      |                                                        |
|    | TRZ | AC     | Transfer if S(I) = A.                                  |
|    | ADD | A      |                                                        |
|    | SUB | K      |                                                        |
|    | TRZ | AC     | Transfer if S(I) = K.                                  |
|    | ADD | K      |                                                        |
|    | SUB | E      |                                                        |
|    | TRZ | AC     | No transfer if S(I) is a propositional variable.       |
|    | CLA | I      |                                                        |
|    | SUB | NUM    |                                                        |
|    | TRZ | AD     | Transfer to AD if I = NUM (initial condition).         |
|    | CLA | I      |                                                        |
|    | ADD | ONE    | ONE is the location of the constant "1."               |
|    | STO | J      |                                                        |
|    | CLA | SUM(J) |                                                        |
|    | ADD | ONE    |                                                        |
| AE | STO | SUM(I) |                                                        |
|    | CLA | ONE    |                                                        |
|    | SUB | I      |                                                        |
|    | TRZ | AF     | Transfer to the ending procedure.                      |
|    | CLA | SUM(I) |                                                        |

|     | TRP | AG        | Transfer if the partial sum is acceptable. |
|-----|-----|-----------|--------------------------------------------|
| AH  | HLT |           | This is the exit if the formula is not well-formed. |
| AB  | CLA | I         | Enter this instruction if S(I) = N. |
|     | SUB | NUM       |  |
|     | TRZ | AH        | Transfer if the last symbol is N. |
|     | CLA | I         |  |
|     | ADD | ONE       |  |
|     | STO | J         |  |
|     | CLA | SUM(J)    |  |
|     | TRA | AE        |  |
| AC  | CLA | I         | Enter this instruction if S(I) is C, A, K, or E. |
|     | SUB | NUM       |  |
|     | TRZ | AH        | Transfer if the last symbol is C, A, K, or E. |
|     | CLA | I         |  |
|     | ADD | ONE       |  |
|     | STO | J         |  |
|     | CLA | SUM(J)    |  |
|     | SUB | ONE       |  |
|     | TRA | AE        |  |
| AD  | CLA | ONE       |  |
|     | TRA | AE        |  |
| AF  | CLA | SUM(ONE)  | Ending procedure. |
|     | SUB | ONE       |  |
|     | TRZ | AI        |  |
|     | TRA | AH        |  |
| AG  | CLA | I         |  |
|     | SUB | ONE       |  |
|     | STO | I         |  |
|     | TRA | AA        |  |
| AI  | HLT |           | This is the exit if the formula is well-formed. |

In examining this program the reader will note that some improvement could be made if we had included some other instructions, such as a transfer on negative, and a transfer on non-zero. Also we need to have print instructions in order to obtain the messages shown in the flowchart, and by use of subscripts as in SUM(I) ($\Sigma_i$ on the flowchart) we have assumed that the computer has the ability to locate information so addressed.

*Example 6.3.* The problem is to sort a series of N words into alphabetical order. We assume that the coding is such that word X precedes word Y if and only if as numbers in the computer X is less than Y. The sorting procedure used is an interchange. The first word is compared to all of the others. Whenever a word which precedes it is found an interchange is made, and the search continues with the new first word. In this way the first word alphabetically becomes the first in the series. The procedure is then continued with the second, third, ... words. The words are in locations W(1) through W(N), and N is stored, as is the constant ONE (1).

```
        CLA  ONE
        STO  I
AF      ADD  ONE
        STO  J
AA      CLA  W(I)
        SUB  W(J)
        TRP  AB      Transfer to interchange words W(I) and W(J).
AD      CLA  J
        ADD  ONE
        STO  J
        SUB  N
        TRP  AC      Transfer if W(I) has been compared with all other
                     words.
        TRA  AA
AB      CLA  W(J)    Interchange W(I) and W(J).
        LDQ  W(I)
        STO  W(I)
        STQ  W(J)
        TRA  AD
AC      CLA  I
        ADD  ONE
        STO  I
        SUB  N
        TRP  AE      Transfer out at the end.
        CLA  I
        TRA  AF
AE      HLT
```

### EXERCISES

Using the digital computer language of this section, write programs (algorithms) to solve the following problems. You may assume in each problem that the data are in storage, and the answer is to be left there.

1. Find the greatest common divisor of two positive integers $a$ and $b$.
   (That is, the largest integer which is a divisor of both $a$ and $b$.)
2. Assuming that we have pennies, nickels, dimes, quarters, and half

dollars, find the least number of coins totaling to a given amount under $10.00.

3. Compute income tax. Assume that a person takes a standard 10% deduction from his gross salary, plus a $600 deduction for each member of his family to obtain his taxable income. Use a tax rate of 20% on the first $5000 of taxable income, 25% on the next $5000, and 30% above that.

4. Program a strategy allowing the first player to win or draw at the game of tic-tac-toe.

## 7. PROGRAMMING LANGUAGES

In the last section we gave three examples of algorithms or programs written for a digital computer. Turing machines or Markov algorithms could be written to accomplish these same tasks, but they would take many pages, and a person writing them would probably make many errors before writing a correct version. Even the computer language which we used in the last section is primitive enough in that people are quite prone to make mistakes in writing long programs. In an effort to reduce the number of errors and to provide programs which can be more readily understood by a person looking at them, a number of programming languages have arisen, and with them translators or algorithms which enable the computer to "read" the languages. Among the better known of these are FORTRAN, ALGOL, MAD, and JOVIAL which are intended primarily for scientific calculations; COBOL for business calculations; IPL-V, LISP, COMIT, and SNOBOL for the processing of lists of information; and most recently PL/1 which has some of the features of all of the above languages.

Although it is not the purpose of this book to teach programming, it is useful for purposes of comparison to see examples of algorithms written in a programming language. For this we have chosen to use the MAD (Michigan Algorithm Decoder) language. The examples presented are the same as those used in the previous section, and the programs are easily readable with just the following comments.

1. An asterisk (*) is used to denote multiplication.

2. The statement "DIMENSION S(100), SUM(100)" in Example 7.2 is used to assign 100 storage locations to the symbols (S) and the partial sums (SUM). This amount could be changed by just changing the number.

3. The statement "NORMAL MODE IS INTEGER" in Example 7.2 indicates that the numerical values in the program are not fractional.

4. Symbols enclosed by dollar signs are interpreted literally, rather than as a name of some quantity.

5. The symbols .E., .NE., .LE., .G. are the relations equal to, not equal to, less than or equal to, and greater than, respectively.

6. The instruction "THROUGH AA, FOR I = N, − 1, I.L.1" means that the sequence of instructions through (and including) the one labeled AA is to be repeated for different values of I. The value of I used initially is N, and this value is changed each time by adding − 1 (subtracting 1) until it becomes less than 1. At that point the instruction following the one labeled AA is executed, and the program continues from there.

· 7. A statement of the form WHENEVER $\mathscr{C}$, $\mathscr{D}$ causes the execution of instruction $\mathscr{D}$ whenever condition $\mathscr{C}$ is fulfilled. Similarly, a statement WHENEVER $\mathscr{C}$ causes the execution of the statements following it up to a statement OR WHENEVER $\mathscr{C}'$, or OTHERWISE, or END OF CONDITIONAL whenever the condition is fulfilled, and the omission of these instructions whenever the condition is not fulfilled.

8. The statement "1 S(I).E.$E$, TRANSFER TO AC" (Example 7.2) is a continuation of the statement on the preceding line, the "1" indicating this.

9. The statement "SYSTEM." is a conventional method of halting the program.

These programs are not necessarily the best which could have been written in MAD to solve these problems, but were chosen because of their clarity.

*Example 7.1.* The solution of two linear equations (see Example 6.1).
```
READ DATA A, B, C, D, E, F
DENOM = A * E − B * D
X = (C * E − B * F)/DENOM
Y = (A * F − C * D)/DENOM
PRINT RESULTS X, Y
END OF PROGRAM
```
*Example 7.2.* Well-formed formulas (see Example 6.2).
```
DIMENSION S(100), SUM(100)
NORMAL MODE IS INTEGER
READ DATA N, S(1)...S(N)
THROUGH AA, FOR I = N, − 1, I.L.1
WHENEVER S(I).E.$N$, TRANSFER TO AB
WHENEVER S(I).E.$C$ .OR. S(I).E.$A$ .OR. S(I).E.$K$ .OR.
1 S(I).E.$E$, TRANSFER TO AC
WHENEVER I.E.N
    SUM(N) = 1
OTHERWISE
    SUM(I) = SUM(I + 1) + 1
```

```
       END OF CONDITIONAL
AE     WHENEVER I.E.1
           WHENEVER SUM(1).NE.1, TRANSFER TO AH
           PRINT COMMENT $ THIS IS A WFF.$
           SYSTEM.
       END OF CONDITIONAL
AA     WHENEVER SUM(I).LE.0, TRANSFER TO AH
AB     WHENEVER I.E.N, TRANSFER TO AH
       SUM(I) = SUM(I + 1)
       TRANSFER TO AE
AC     WHENEVER I.E.N, TRANSFER TO AH
       SUM(I) = SUM(I + 1) − 1
       TRANSFER TO AE
AH     PRINT COMMENT $ THIS IS NOT A WFF.$
       END OF PROGRAM
```

*Example 7.3.* A simple sort. (See Example 6.3.)

```
       DIMENSION W(100)
       NORMAL MODE IS INTEGER
       READ DATA N, W(1)...W(N)
       THROUGH AA, FOR I = 1, 1, I.E.N
       THROUGH AA, FOR J = I, 1, J.G.N
       WHENEVER W(I).G.W(J)
           TEMP = W(I)
           W(I) = W(J)
           W(J) = TEMP
AA     END OF CONDITIONAL
       PRINT RESULTS W(1)...W(N)
       END OF PROGRAM
```

## References

1. Arden, B., B. A. Galler, and R. Graham, *The Michigan Algorithm Decoder.* Edwards Bros., Ann Arbor, Michigan, 1965.
2. Davis, Martin. *Computability and Unsolvability.* McGraw-Hill Book Company, New York, 1958.
3. Davis, Martin, ed. *The Undecidable.* Raven Press, Hewlett, New York, 1965. (This contains Turing's paper [19] and other basic papers.)
4. Farber, D. J., R. E. Griswold, and I. P. Polonsky. "SNOBOL, a string manipulation language." *Journal of the Association for Computing Machinery* **11** (1964), 21–30.
5. Galler, B. A. *The Language of Computers.* McGraw-Hill Book Company, New York, 1962.
6. Gill, A. *Introduction to the Theory of Finite-State Machines.* McGraw-Hill Book Company, New York, 1962.

7. Ginsburg, S. *An Introduction to Mathematical Machine Theory.* Addison-Wesley Publishing Company, Reading, Mass., 1962.

8. Levin, M. I. *LISP 1.5 Programmer's Manual.* MIT Press, Cambridge, Mass., 1962.

9. Lin, S. and T. Rado. "Computer studies of Turing machine problems." *Journal of the Association for Computing Machinery* **12** (1965), 196–212.

10. McCracken, D. D. *A Guide to FORTRAN IV Programming.* John Wiley and Sons, New York, 1965.

11. McCracken, D. D. *A Guide to ALGOL Programming.* John Wiley and Sons, New York, 1962.

12. McCracken, D. D. *A Guide to COBOL Programming.* John Wiley and Sons, New York, 1963.

13. Markov, A. A. *Theory of Algorithms.* Israel Program for Scientific Translations, Jerusalem, 1961.

14. Moore, E. F., ed. *Sequential Machines.* Addison-Wesley Publishing Company, Reading, Mass., 1964.

15. Newell, A., ed. *Information Processing Language-V Manual.* Prentice-Hall, Englewood Cliffs, New Jersey, 1961.

16. Rado, T. "On non-computable functions." *Bell System Technical Journal* **41** (1962), 877–884.

17. Shaw, C. J. *A Programmer's Introduction to Basic JOVIAL.* System Development Corporation, Santa Monica, California, TM-629, 1961.

18. Trakhtenbrot, B. A. *Algorithms and Automatic Computing Machines.* D. C. Heath and Company, Boston, Mass., 1963.

19. Turing, A. M. "On computable numbers, with an application to the Entscheidungs-problem." *Proceedings of the London Mathematical Society* (2), **42** (1936–1937), 230–265.

20. Yngve, V. H. *An Introduction to COMIT Programming.* MIT Press, Cambridge, Mass., 1962.

# 6. The First-Order Predicate Calculus

In Chapter 3 we developed a system of logic which may be used to deal with propositions, that is, with declarative sentences which have a fixed truth value—either "true" or "false." We now wish to examine these propositions in more detail, and to expand our logical system so that we may take into account some of the inner structure of the propositions. For example, while we can handle the sentence "If some cheese is green then all cheese is green" in the form $p \supset q$, we would like to have a logic which is capable of expressing the fact that the $p$ and $q$ are both about green cheese.

In particular, we wish to be able to work with individuals, relations between individuals, and the properties of sets of individuals. For example, we may wish to describe individuals called "Henry" and "John," and to talk about the fact than Henry is John's father. Thus we have both functions on the set of individuals (for example, defining the father of a given individual), and predicates describing properties and relations on sets of individuals (for example, $P$ may denote the binary relation "is a father of"). The functions define new individuals in terms of previously known ones, and the predicates by their truth values describe whether a set of individuals has a certain relationship or property.

At the same time, we wish our logic to be sufficiently strong to deal with *sentence forms*—structures which appear as declarative sentences, but which have no definite truth value because of the appearance of *individual variables*. Thus "$2 + 3 = 4$" is a proposition which has a truth value, but "$x + 3 = 4$" is a form which is neither true nor false since we do not know which number is represented by $x$. Similarly, "Jefferson Davis was a student at Purdue" is either true or false (although it might take some time to determine which), but "he was a student at Purdue" is neither true nor false since we do not know to whom "he" refers. The logical system which we shall describe is called the *first-order predicate calculus*.

## 1. DEFINITIONS AND BASIC PROPERTIES

The logic which we shall develop will be an extension of our previous system, and thus we shall make use of our previous symbols for the logical operations and for propositions. In addition, we shall need the following symbols:

for *individual constants* (names of individuals):
$a, b, c, a_1, \ldots$
for *individual variables* (pronouns): $x, y, z, x_1, \ldots$
for *function letters* (functions): $f_j^i, g_j^i, h_j^i, \ldots$, where $i$ and $j$ denote positive integers
for *predicate letters* (predicates): $F_j^i, G_j^i, H_j^i, \ldots$, where $i$ and $j$ denote positive integers
for *quantifiers* (whose function will be explained below) $(\alpha), (\exists \alpha)$, where $\alpha$ may be any individual variable.

With the introduction of these new symbols, our previous concepts need extension and redefinition.

**Definition 1.1.** A *term* is defined as follows:

1. Individual constants and individual variables are terms.
2. If $f_j^n$ is a function letter and $t_1, t_2, \ldots, t_n$ are terms, then $f_j^n(t_1, \ldots, t_n)$ is a term.
3. The only terms are those formed by (1) and (2).

**Definition 1.2.** A string is an *atomic formula* if it is either

1. a propositional variable standing alone, or
2. a string of the form $F_j^n(t_1, \ldots, t_n)$, where $F_j^n$ is a predicate letter and $t_1, \ldots, t_n$ are terms.

Note that in this definition the superscript on the function or predicate letter corresponds to number of terms $t_1, \ldots, t_n$. A function or predicate letter whose superscript is $n$ is termed *n-ary*. In each case, the subscript is just an index to differentiate between various predicates or functions. Thus $F_1^1$ and $F_2^1$ are two different predicates or properties of one individual; while $F_1^1$ and $F_1^2$ are also two different predicates, but one of these $(F_1^1)$ is a property of one individual, and the other $(F_1^2)$ is a property of a pair of individuals. In general, we shall omit the superscripts whenever the context makes the number of arguments clear, and use distinct function and predicate letters in order to avoid the necessity of using subscripts.

As symbols in the metalanguage we shall use script capital letters to denote formulas, and lower case Greek letters to denote individual constants and variables.

**Definition 1.3.** A *well-formed formula* (*wff*) is defined as follows:

1. An atomic formula is a wff.
2. If $\mathscr{A}$ is a wff and $\alpha$ is an individual variable, then $(\alpha)\mathscr{A}$ and $(\exists \alpha)\mathscr{A}$ are wffs.

3. If $\mathscr{A}$ and $\mathscr{B}$ are wffs, then $\sim(\mathscr{A})$, $(\mathscr{A}) \supset (\mathscr{B})$, $(\mathscr{A}) \wedge (\mathscr{B})$, $(\mathscr{A}) \vee (\mathscr{B})$, and $(\mathscr{A}) \equiv (\mathscr{B})$ are wffs.

4. The only wffs are those obtainable by finitely many applications of (1), (2), and (3).

Notice here that in (2), $\mathscr{A}$ may be any wff, and $\alpha$ any individual variable. Thus $(\alpha)\mathscr{A}$ and $(\exists\alpha)\mathscr{A}$ are both wffs, whether or not $\alpha$ actually occurs in $\mathscr{A}$.

We shall make use of the same parenthesis, heirarchy, and dot conventions as were adopted in Chapter 3. The quantifiers will occupy the same hierarchy level as the negation sign. In particular, $(x)(y)\mathscr{A}$ is an abbreviated form of the wff $(x)((y)(\mathscr{A}))$; and similarly for the other pairs and sequences of quantifiers.

The new symbols which we have formally introduced require interpretation. We are concerned here with a set or *domain* of individuals. The individual variables and the individual constants represent elements of this set, the distinction being that the constants represent specific elements, whereas the variables do not. Thus the domain might be the set of natural numbers, with the individual constants being $0, 1, \ldots$ Or the domain might be the employees of a firm, with the individual constants denoting the president, personnel manager, and others.

The function letters represent functions or maps from product sets of the domain into the domain. For example, if the domain is the set of natural numbers, one function might assign to every pair of numbers their greatest common divisor. If the domain is the set of employees of a firm, a function might assign to each employee his immediate supervisor. Similarly, the predicate letters represent predicates on product sets of the domain. Thus they correspond to maps into the set $\{T, F\}$.

The two quantifiers, ( ) and $(\exists)$, are called respectively the *universal* and the *existential* quantifiers, and are the formal equivalents of the words "all" and "some." Thus if $P$ (that is, $P_1^1$) is a predicate letter and $x$ is an individual variable, $(x)P(x)$ may be read "for all $x$ in the domain, the individual $x$ has the property $P$," and $(\exists x)P(x)$ may be read "there exists an (some) individual $x$ in the domain, which has the property $P$." These two quantifiers are not independent, but are in fact related by the logical equivalence

$$(x)P(x) \text{ eq } \sim(\exists x) \sim P(x)$$

As was mentioned at the beginning of this chapter, a sentence form or propositional form does not have a well-defined truth value because of the presence of variables in it. However, we can easily think of statements which ostensibly contain variables and yet are propositions rather than propositional forms. If we examine these, we find that the situation which most frequently causes them to be propositions and to have truth values is one in which the informal equivalents of our quantifiers are present. Think for the moment of the domain of individuals in a particular room, and of the English sentences, "He went to Purdue," "Someone in this room went to Purdue," "Everyone in this room went to Purdue." Symbolically we represent these sentences by $P(x)$, $(\exists x)P(x)$, and $(x)P(x)$; and we recognize that the first is a propositional form, while the remaining two are propositions and have truth values. Thus, informally at least, quantifiers transform propositional forms into propositions whose truth values we can determine and discuss.

## EXERCISES

1. Determine whether each of the following is (a) a term, (b) an atomic formula, (c) a non-atomic wff, (d) none of these.
   (a)  $(x)(A(x, y) \supset (D(y) \lor f(x, a)))$
   (b)  $f(x, a)$
   (c)  $(x)(y)D(z, a)$
   (d)  $A(a, b) \lor B(x, a)$
   (e)  $B(a, f(x, a))$
   (f)  $(x)C(x, y, a)$
   (g)  $a$
   (h)  $h(a)$
   (i)  $(x)A(f(x, y), b) \land B(x, a)$
   (j)  $A(x, a, y)$
   (k)  $g(x, f(x, y), a, g(a, b, x, y))$
   (l)  $x \equiv y$
   (m)  $D(a, b, c)$
   (n)  $f(x, A(x, y))$
   (o)  $(x)x$
   (p)  $\sim(\exists x) \sim f(x)$
   (q)  $A(x) \supset \sim\sim(\exists x) \sim B(x, y)$
2. Write formulas of the first-order predicate calculus corresponding to each of the following sentences.
   (a) Every green cheese is a hamburger.
   (b) If some widget flacculates, then there is a bug-eyed monster chasing that widget.
   (c) Anyone who is eligible is either a descendant of the Cabots and the Lodges, or of the Lodges and the Lowells.
   (d) For every boy present there will be at least one girl present.
   (e) For every boy present there will be exactly one girl present.

3. Interpret each of the following formulas in reasonably good English.
   (a) $(x)(\exists y)(A(x, y) \supset (z)B(x, y, z))$
   (b) $(x)(y)(A(x, y) \vee (F(x) \wedge G(f(x, y))))$
   (c) $\sim(\exists x)A(x, y) \vee (y)A(y, x)$
   (d) $(x)(y)(E(x, y) \equiv E(y, x))$
   (e) $(x)(y)(A(x, y) \wedge (\exists z)(A(x, z) \vee (\exists x)A(x, y)))$

## 2. FREE AND BOUND VARIABLES; SUBSTITUTION

We are led then to the distinction between variables which are quantified and those which are not, and in general to the question of which quantifier in an expression controls which variables. We speak of the expression to which the quantifier is applied as the *scope* of the quantifier, and speak of an occurrence of an individual variable $x$ as being *bound* if it is either an occurrence $(x)$ or $(\exists x)$, or within the scope of a quantifier $(x)$ or $(\exists x)$. Any other occurrence of a variable is a *free* occurrence. Let us consider several examples in amplification of these ideas.

*Example 2.1.* $(x)A(x) \supset B(y)$. In this expression the scope of the quantifier is the expression $A(x)$, and thus the $x$ is a bound variable in both of its occurrences. The variable $y$ occurs free, since it is neither within the scope of a quantifier on $y$, nor is in a quantifier $(y)$ or $(\exists y)$.

*Example 2.2.* $(x)(A(x) \supset B(y))$. Here the scope of the quantifier is the expression $A(x) \supset B(y)$, and $x$ again occurs as a bound variable. Even though the occurrence of $y$ is now within the scope of a quantifier, the quantifier is not on $y$, so $y$ is still free.

*Example 2.3.* $(\exists x)(A(x) \vee (y)B(x, y))$. We have here two quantifiers. The first, $(\exists x)$, has as its scope the expression $A(x) \vee (y)B(x, y)$, and thus all three occurrences of $x$ (in $(\exists x)$, in $A(x)$, and in $B(x, y)$) are bound. The second quantifier, $(y)$, has as its scope the expression $B(x, y)$, and thus $y$ is also a bound variable in both of its occurrences.

*Example 2.4.* $A(x) \wedge (x)(B(x) \equiv (\exists y)C(y))$. In this expression, the first occurrence of $x$ is obviously free since it is not within the scope of any quantifier, while the second and third occurrences of $x$ and the occurrences of $y$ are all bound. Thus a variable may have both free and bound occurrences within a single expression.

*Example 2.5.* $A(x) \wedge (x)(B(x) \equiv (\exists x)C(x))$. In this expression the first occurrence of $x$ is again free. The third occurrence of $x$ (in $B(x)$) is within the scope of the quantifier $(x)$ but not within the scope of the existential quantifier, so it is bound by the universal quantifier. However, the fifth occurrence of $x$ (in $C(x)$) is within the scope of both quantifiers on $x$, and thus there might be some question as to which quantifier is applicable. (Clearly this occurrence of $x$ is bound.) The rule is this: *an occurrence of an individual variable is bound by the innermost quantifier on that variable within whose scope that particular occurrence lies.* Thus in the present example, the $x$ in $C(x)$ is bound by the existential rather than by the universal quantifier.

One might question why such an expression as that in Example 2.5 is allowed in our system. The $x$ in $C(x)$ has nothing to do with the universal quantifier: why not use the expression in Example 2.4 instead, since this appears to be equivalent? And in fact, some systems do have just this sort of rule. However, such a rule complicates the definition of wffs, and causes the process of substitution, which we shall presently discuss, to be burdened with much checking to insure that all variables are kept distinct.

In the propositional calculus we could pass lightly over substitution, for it causes no difficulties. However, with the distinction between free and bound occurrences of a variable, substitution is not so lightly undertaken. We do not wish to go into all of the ramifications here, so we restrict ourselves to three remarks. First, substitution of terms for individual variables is generally restricted to the free occurrences. We may make a change of variables for the bound occurrences (for example, change the expression of Example 2.4 to that of Example 2.5), but that is about all; such a change is independent of what is done with the free occurrences. Second, as in the propositional calculus, we must substitute either for all free occurrences of a given variable, or for none. Third, care must be taken so that a free variable does not become bound by the substitution, as this changes the meaning of an expression.

*Example 2.6.* $(y)(x = y)$. First, we have an expression here which does not seem to fit our definition of wff. But this is only apparently a fault, for we could have written the expression as $(y)E(x, y)$ where the predicate letter $E$ denotes equality. However, we chose to use the more familiar notation. In this expression $x$ occurs free and $y$ occurs bound. Thus substitutions other than a change of variable are restricted to the $x$. The expression which we have is a sentence form and as such is neither true nor false. However, we can say that if the domain of individuals contains at least two elements, then no matter what value is assigned to $x$, the resulting proposition is not true. For example, if the domain is the set $\{a, b\}$ and $x$ is assigned the value $a$, the given sentence form becomes the sentence $(y)(a = y)$; for the given domain this is equivalent to the sentence $(a = a) \land (a = b)$. This sentence is not true if we interpret equality in the usual sense. A similar analysis holds if $x$ is assigned the value $b$.

Now, if we substitute $z$ for $x$ in this sentence form, we obtain the form $(y)(z = y)$. Some cogitation shows that the above analysis applies equally well to this form: no matter what value $z$ is assigned from a domain of at least two individuals, the resulting sentence is not true.

This all changes if we substitute $y$ for $x$: for the resulting *sentence* is $(y)(y = y)$, which is clearly true for all domains with the usual interpretation of equality. More fundamentally, note that we no longer have a sentence form, but a sentence.

Thus we must include in any substitution rules clauses which prevent this sort of change of free variables to bound variables.

EXERCISES

1. In each of the following, identify the free and bound occurrences of each variable.
   (a) $(x)A(x) \supset (\exists y)B(x, y)$
   (b) $A(x, y) \wedge (\exists x)B(y) \supset (y)(z)C(x, y, z)$
   (c) $(x)(\exists y)(A(y, x) \wedge (y)C(y)) \supset B(x, y)$
   (d) $(x)(y)(A(z) \supset B(z))$
   (e) $A(x) \supset (B(y) \supset (\exists x)(C(y) \supset (y)D(x)))$
2. In each of the following, perform the indicated substitutions in the corresponding formulas of Exercise 1, if the substitutions are legal.
   (a) In the formula of 1(a), substitute $f(x, z)$ for $x$.
   (b) Substitute $z$ for $x$, and $g(y, z)$ for $y$.
   (c) Substitute $y$ for $x$, and $f(x, y)$ for $y$.
   (d) Substitute $x$ for $z$.
   (e) Substitute $f(y)$ for $x$, and $f(y)$ for $y$.

## 3. VALIDITY AND SATISFIABILITY

We now wish to consider the notions of truth and falsity in the first-order predicate calculus. Because of the presence of the domain and the predicate and function letters, the situation is more complex here than in the propositional calculus. Throughout this section we will consider three formulas: $\mathscr{F}_1$: $F(f(x, a), b)$; $\mathscr{F}_2$: $(\exists x)F(f(x, a), b)$; $\mathscr{F}_3$: $(x)F(f(x, a), b)$. Each of these formulas has a predicate letter $(F)$, a functional letter $(f)$, an individual variable $(x)$, and two individual constants ($a$ and $b$). In $\mathscr{F}_1$ the individual variable is free, so that we have a sentence form: the idea of truth must depend on assignments of individuals to this variable. However, $\mathscr{F}_2$ and $\mathscr{F}_3$ contain no free individual variables, and hence are propositions. Thus they have a truth value. Since there are no propositional connectives in these formulas, in the propositional calculus $\mathscr{F}_2$ (and $\mathscr{F}_3$) would be represented by a single propositional letter, and we could do no better than assert that $\mathscr{F}_2$ (or $\mathscr{F}_3$) takes on either the value T or the value F. But in the first-order predicate calculus we are able to make assignments to each of the individual constants and variables, and to the predicate and function letters, and thus to analyze the situation more deeply.

*Example 3.1.* Let us consider $\mathscr{F}_2$. Assignments of values to the individual constants and variables must be made from some domain, the distinction being that the assignments to the constants are considered fixed, and the assignments to the variables are allowed to range over the entire domain. We must also assign meaning to the predicate and function letters: they represent certain relations and mappings. For example, we might take the domain to be the natural numbers, and interpret $F$ as the equality predicate, $f$ as addition, $a$ as the integer 2, and $b$ as 5. Then using more conventional notation, $\mathscr{F}_2$ is the formula $(\exists x)(x + 2 = 5)$. With the chosen domain this formula has the value T; but with another domain such as $\{2\} \cup \{x \mid x = 5y \text{ for some integer } y\}$, it has the value F. Other assignments may be made for this formula. To cite one more, we choose as the domain the set of letters of the English alphabet, let $f(\alpha, \beta)$ be that one of $\alpha$ and $\beta$ which occurs first in the alphabet, and let $F(\alpha, \beta)$ be the predicate "$\alpha$ precedes $\beta$ in alphabetic order."

**Definition 3.1.** Given a wff $\mathscr{F}$ of the first-order predicate calculus, an *interpretation* of $\mathscr{F}$ consists of a non-empty domain $D$ and an assignment to each $n$-ary predicate letter of an $n$-ary predicate on $D$, to each $n$-ary function letter of an $n$-ary function on $D$, and to each individual constant of a fixed element of $D$.

**Definition 3.2.** A wff of the first-order predicate calculus is *satisfiable in a domain D* if there exist an interpretation with domain $D$, and assignments of elements of $D$ to the free occurrences of individual variables in the formula such that the resulting proposition is true. A wff is *valid in a domain D* if for every interpretation with domain $D$ and every assignment of elements of $D$ to the free occurrences of individual variables in the formula the resulting proposition is true. A wff is *satisfiable* if it is satisfiable in some domain; it is *valid* if it is valid in all domains.

We will also speak of a formula being valid or satisfiable in a particular interpretation.

Certain relationships between satisfiability, validity, and the domain are easy to establish. If a wff $\mathscr{F}$ is valid in a domain $D$, then it is satisfiable in $D$. Similarly, if $\mathscr{F}$ is valid, then it is satisfiable. If $\mathscr{F}$ is valid in $D$, then it is valid in any nonempty subset of $D$; if $\mathscr{F}$ is satisfiable in $D$, then it is satisfiable in any domain which contains $D$ as a subset. $\mathscr{F}$ is valid if and only if $\sim\!\mathscr{F}$ is not satisfiable; $\mathscr{F}$ is satisfiable if and only if $\sim\!\mathscr{F}$ is not valid. This indicates that the duality which holds for Boolean algebras and the propositional calculus is extendible to the first-order predicate calculus.

More important than these relationships is that expressed by the *Löwenheim Theorem*: if a wff is satisfiable in some domain it is satisfiable in the domain of natural numbers; and its dual: if a wff is valid in the domain of natural numbers, it is valid in any domain. Together these theorems imply that in any discussion of validity and satisfiability for the first-order predicate calculus, it is possible to restrict the domains to the natural numbers and subsets of these.

*Example 3.2.*  Let us consider the three formulas $\mathscr{F}_1$, $\mathscr{F}_2$, and $\mathscr{F}_3$ given at the beginning of this section, and take three domains defined as follows: $D_1 = \{2, 3, 5\}$; $D_2 = \{x \mid$ for some non-negative integers $y$ and $z$, $x = 2y + 5z\}$; $D_3 = $ the set of all positive integers. We take $F$ to be equality, and $f$ to be ordinary addition in $D_2$ and $D_3$. In $D_1$ we take $f$ as an "addition" defined by Table 3.1. In each domain we take $a$ to be 2 and

**Table 3.1**

$f(x, y)$ for $D_1$

| $x \backslash y$ | 2 | 3 | 5 |
|---|---|---|---|
| 2 | 5 | 5 | 3 |
| 3 | 5 | 3 | 2 |
| 5 | 5 | 2 | 5 |

$b$ to be 5. The proper statements of validity and satisfiability for these interpretations and in general are given in Table 3.2.

**Table 3.2**

Validity and satisfiability with the given interpretations

| Formula | $D_1$ | $D_2$ | $D_3$ | Valid? | Satisfiable? |
|---|---|---|---|---|---|
| $\mathscr{F}_1$ | Valid | Not satisfiable | Satisfiable | No | Yes |
| $\mathscr{F}_2$ | Valid | Not satisfiable | Valid | No | Yes |
| $\mathscr{F}_3$ | Valid | Not satisfiable | Not satisfiable | No | Yes |

### EXERCISES

1. Consider the formulas $\mathscr{F}_1$, $\mathscr{F}_2$, and $\mathscr{F}_3$, the domain $\{a, b, \ldots, z\}$, and the assignments for $F$ and $f$ stated at the end of Example 3.1. For each

of these formulas, answer the following question: what assignments of elements from the domain may be made to the individual constants $a$ and $b$ so that the formula is valid in the given interpretation? (Assume that $f(\alpha, \alpha) = \alpha$, and that $F(\alpha, \alpha)$ is true.)

2. Consider the two formulas

$$(x)(\exists y)(A(x) \supset B(y, f(x)))$$
$$(\exists x)((y)B(y, f(x)) \supset A(x))$$

Determine whether each of these formulas is true or false in the following interpretations.

(a) Domain: the integers; $A(\alpha)$: $\alpha$ is a positive integer; $B(\alpha, \beta)$: $\alpha = \beta$; $f(\alpha) = \alpha^2$.

(b) Domain: the integers; $A(\alpha)$: $\alpha$ is an even integer; $B(\alpha, \beta)$: $\alpha$ is greater than $\beta$; $f(\alpha) = \alpha + 1$.

(c) Domain: English words; $A(\alpha)$: $\alpha$ is a noun; $B(\alpha, \beta)$: $\alpha$ and $\beta$ are the same word; $f(\alpha) = $ the plural of $\alpha$.

(d) Domain: strings in $\{a, b, \ldots, z\}$; $A(\alpha)$: $\alpha$ is an English adjective; $B(\alpha, \beta)$: $\alpha$ is the same string as $\beta$; $f(\alpha) = $ the string formed by appending "est" to the right of the string $\alpha$.

(e) Domain: people; $A(\alpha)$: $\alpha$ is a father; $B(\alpha, \beta)$: $\alpha$ is the mother of $\beta$; $f(\alpha) = $ the oldest child of $\alpha$.

(f) Domain: people; $A(\alpha)$: $\alpha$ was murdered; $B(\alpha, \beta)$: $\alpha$ arrested $\beta$; $f(\alpha) = $ the murderer of $\alpha$.

## 4. THE DETERMINATION OF TRUTH VALUES

Two facts should be clear from the discussion in Section 3. First, the valid propositions and propositional forms constitute the extension of the concept of a tautology; and second, the existence of various domains and assignments to the predicate and function letters greatly complicates the truth tables. Let us examine the wff

$$P(x) \wedge (y)Q(x, y) \supset .Q(x, z) \tag{4.1}$$

as an example of the techniques involved in the determination of truth values.

Let us assume that the domain under consideration contains exactly two elements, $a$ and $b$. We note first that there are two free variables ($x$ and $z$) and two predicate letters ($P$ and $Q$). Beginning with $P$, we inquire as to which predicates $P$ may represent. We are not concerned with all of the possible ways in which to state these predicates, but only with the predicates which correspond to distinct mappings into the set $\{T, F\}$. Because there is only one variable involved which may take on either of the two values (the elements of the domain), $P$ may represent any one of four functions as shown in Table 4.1. Similarly, since $Q$ is a binary

**Table 4.1**

Assignments to $P$

| $x$ | $\lambda_1(x)$ | $\lambda_2(x)$ | $\lambda_3(x)$ | $\lambda_4(x)$ |
|---|---|---|---|---|
| $a$ | T | T | F | F |
| $b$ | T | F | T | F |

predicate, it may represent any one of sixteen functions as indicated in Table 4.2.

**Table 4.2**

Assignments to $Q$

| $u$ | $v$ | $\theta_1(u, v)$ | $\theta_2(u, v)$ | $\ldots$ | $\theta_{16}(u, v)$ |
|---|---|---|---|---|---|
| $a$ | $a$ | T | T | | F |
| $a$ | $b$ | T | T | | F |
| $b$ | $a$ | T | T | | F |
| $b$ | $b$ | T | F | | F |

In order to assign a truth value to the given statement form, we must make four assignments: to $x$, $z$, $P$, and $Q$. Since these assignments may be made in 2, 2, 4, and 16 ways respectively, the truth table, Table 4.3, has 256 lines.

To examine a particular line in the table, let us consider the assignments $x = b$, $z = a$, $P = \lambda_3$, and $Q = \theta_{11}$. The statement then becomes

$$\lambda_3(b) \wedge (y)\theta_{11}(b, y) \supset .\theta_{11}(b, a)$$

which has the truth value T since, although $\theta_{11}(b, a)$ is false, also, so is $(y)\theta_{11}(b, y)$.

If we examine each line of this truth table, we will find nothing but T for the truth value. Thus this particular statement is valid in any domain of two individuals. However, this analysis does *not* guarantee that the statement is (universally) valid: this would require the examination of the truth table for the domain of natural numbers with all possible interpretations of $P$ and $Q$—clearly a formidable task.

Can we avoid such a task? That is, do other techniques exist besides the truth table technique for determining validity? Is it possible to determine whether or not a given statement is valid? If we examine our example again, we see quite easily that it is in fact a valid statement. In any domain the statement will take on the value F if and only if $P(x)$ has the value T, $(y)Q(x, y)$ has the value T, and $Q(x, z)$ has the value F for a particular assignment. But clearly, if the domain contains individuals $a_1, a_2, a_3, \ldots$, and $(y)Q(x, y)$ has the value T for a given assignment to $x$ and $Q$, then for that assignment the following statements all have the value T: $Q(x, a_1)$, $Q(x, a_2)$, $Q(x, a_3)$, .... Since $z$ must be assigned one of the values $a_1, a_2, a_3, \ldots$, it is impossible in any given domain for $Q(x, z)$ to have the value F while $(y)Q(x, y)$ has the value T. Thus in any given domain the statement is valid, and hence it is (universally) valid.

**Table 4.3**

Truth table for (4.1)

| $x$ | $z$ | $P$ | $Q$ | $P(x)$ | $(y)Q(x, y)$ | $Q(x, z)$ | Statement |
|---|---|---|---|---|---|---|---|
| $a$ | $a$ | $\lambda_1$ | $\theta_1$ | T | T | T | T |
| . . . . . | | | | | | | |
| $a$ | $a$ | $\lambda_1$ | $\theta_{16}$ | T | F | F | T |
| $a$ | $a$ | $\lambda_2$ | $\theta_1$ | T | T | T | T |
| . . . . . | | | | | | | |
| $a$ | $b$ | $\lambda_1$ | $\theta_1$ | T | T | T | T |
| . . . . . | | | | | | | |
| $b$ | $b$ | $\lambda_4$ | $\theta_{16}$ | F | F | F | T |

Here we have a different technique which works—at least for this particular statement. And for most of the statements that we encounter in applications, techniques can be devised which will determine satisfiability or validity. However, it can be shown that there is no algorithm for validity or satisfiability which will work for all statements. For the propositional calculus we have such an algorithm—the truth table technique. Here that fails to be an algorithm because of the infinitely large domain which must generally be considered and the infinitely many assignments which must be made to cover all interpretations.

In other words, although we can devise a technique which will work for a given statement, we can always find a statement which the given technique will fail to analyze. Moreover, it is *not* true that we can devise a technique which will work for any particular statement, regardless of what the statement is. It has been shown that there exist statements which cannot be analyzed by any algorithmic techniques, the so-called *undecidable* statements. In fact, the halting problem for Turing machines and the Busy Beaver problem may be formulated as undecidable statements of the first-order predicate calculus.

<h3 style="text-align:center">EXERCISES</h3>

For each of the following formulas, determine validity or satisfiability in domains of one and two individuals.

1. $(x)(y)(B(x, a) \lor B(a, y))$
2. $A(x) \supset (\exists y)(B(x, y) \land B(y, a))$
3. $(y)(B(f(x), y) \supset A(y))$
4. $(y)B(f(x), y) \supset A(y)$
5. $(x)A(x, a) \supset \sim(B(f(x)) \lor ((\exists y)A(g(x, y), b) \not\equiv C(z)))$

## 5. THE PRENEX NORMAL FORM

As in the propositional calculus, there are several normal or canonical forms which are generally accepted for formulas of the first-order predicate calculus. In this section we wish to discuss the simplest of these.

**Definition 5.1.** A formula of the first-order predicate calculus is in *prenex normal form* if it is of the form $Q_1(x_1)Q_2(x_2)\ldots Q_n(x_n)\mathcal{M}$, where (1) each $Q_i(x_i)$ is either $(x_i)$ or $(\exists x_i)$, (2) $\mathcal{M}$ is a wff, called the *matrix*, involving no quantifiers, (3) $x_1, x_2, \ldots, x_n$ are distinct variables each of which occurs in $\mathcal{M}$, and (4) the scope of $Q_n(x_n)$ is $\mathcal{M}$.

In the above definition no conditions are placed on $\mathcal{M}$ other than that it be quantifier-free. If desired, one can consider $\mathcal{M}$ as a formula of the propositional calculus and impose one of the normal forms from that study on it. Notice that since $\mathcal{M}$ is a wff the entire formula is well-formed.

Since we have shown that the conditional and negation form a minimal set of connectives (Chapter 3, Section 8, Problem 2a), in order to show that each wff of the first-order predicate calculus has an equivalent formula in prenex normal form, it is sufficient to show that the quantifiers may be moved forward across the negation and conditional. For this purpose we adopt the following notation: if $\mathcal{A}$ is a formula containing the individual variable $x$ and we are particularly concerned with $x$, we will write $\mathcal{A}(x)$. Having written $\mathcal{A}(x)$, by $\mathcal{A}(y)$ we shall mean the formula obtained by substituting $y$ for all free occurrences of $x$ in $\mathcal{A}(x)$. Notice that other variables may also occur in $\mathcal{A}(x)$, and in particular, $y$ may occur in it.

With this notation we may then state the algorithm for deriving a formula in prenex normal form which is equivalent to the given one.

1. Replace any well-formed subformula (wfsf) $\sim (x)\mathscr{A}(x)$ by $(\exists x) \sim \mathscr{A}(x)$ (where $x$ is any individual variable).

2. Replace any wfsf $\sim (\exists x)\mathscr{A}(x)$ by $(x) \sim (\mathscr{A})(x)$.

3. Replace any wfsf $(x)\mathscr{A}(x) \supset \mathscr{B}$ by $(\exists y)(\mathscr{A}(y) \supset \mathscr{B})$, where $y$ is any individual variable which does not occur free in either $\mathscr{A}(x)$ or $\mathscr{B}$.

4. Replace any wfsf $(\exists x)\mathscr{A}(x) \supset \mathscr{B}$ by $(y)(\mathscr{A}(y) \supset \mathscr{B})$, where $y$ does not occur free in either $\mathscr{A}(x)$ or $\mathscr{B}$.

5. Replace any wfsf $\mathscr{A} \supset (x)\mathscr{B}(x)$ by $(y)(\mathscr{A} \supset \mathscr{B}(y))$, where $y$ is not free in $\mathscr{A}$ or $\mathscr{B}(x)$.

6. Replace any wfsf $\mathscr{A} \supset (\exists x)\mathscr{B}(x)$ by $(\exists y)(\mathscr{A} \supset \mathscr{B}(y))$, where $y$ is not free in $\mathscr{A}$ or $\mathscr{B}(x)$.

These transformations are reversible, and one may prove that any formulas obtained by applying such transformations to a given formula are logically equivalent to the original formula. By applying these transformations as long as possible, one arrives at a formula in prenex normal form in a finite number of steps. To specify an algorithmic process completely, one would need also to state in which order to apply these transformations, and in which order to choose the replacement variables $y$ in rules 3 to 6. This change of variables is dictated by the requirement that the transformation not bind any variables which are originally free.

*Example 5.1.* Consider the formula

$$(x)A(x, y) \supset \sim (\exists z)(B(x, z) \supset (y)A(x, y))$$

Choosing the variables for substitution in the order $x, y, z, x_1, y_1, \ldots$, and indicating the part of the formula which becomes the matrix by square brackets rather than parentheses, we have the following sequence:

(1) $(\exists z)[A(z, y) \supset \sim (\exists z)(B(x, z) \supset (y)A(x, y))]$

(2) $(\exists z)[A(z, y) \supset (z) \sim (B(x,z) \supset (y)A(x, y))]$

(3) $(\exists z)(x_1)[A(z, y) \supset \sim (B(x, x_1) \supset (y)A(x, y))]$

(4) $(\exists z)(x_1)[A(z, y) \supset \sim (z)(B(x, x_1) \supset A(x, z))]$

(5) $(\exists z)(x_1)[A(z, y) \supset (\exists z) \sim (B(x, x_1) \supset A(x, z))]$

(6) $(\exists z)(x_1)(\exists y_1)[A(z, y) \supset \sim (B(x, x_1) \supset A(x, y_1))]$

*Comments:* (1) • Notice that we could not have retained the $x$ when moving the quantifier outside, since that would have bound the previously free occurrences of $x$ on the right-hand side of the formula. (3) • Similarly, we could not have here retained the $z$ since although the $z$ in $A(z, y)$ is bound by the external quantifier, it is free as far as the formula inside the brackets is concerned. Notice that this procedure automatically takes care of the problem of having two quantifiers on the same variable, in this case $(\exists z)(z)$, over a formula. (6) ● This formula is in prenex normal form. The free occurrences of variables ($x$ and $y$) are the same as in the original formula. The formula within the matrix may now be treated grossly as a formula of the propositional calculus: $p \supset \sim(q \supset r)$, or more finely as a formula of the first-order predicate calculus, taking into account the relationships between the $p$, $q$, and $r$.

*Example 5.2.* We now consider the formula

$$\sim(\exists y)(\sim A(y) \equiv (\exists x)B(y, x))$$

In the absence of rules for moving the quantifiers across other connectives, we must transform the given formula into an equivalent one involving only the conditional and negation. This formula has the propositional form $p \equiv q$, so we use the equivalence $p \equiv q$ eq $\sim((p \supset q) \supset \sim(q \supset p))$. Our formula then becomes

$$\sim(\exists y) \sim((\sim A(y) \supset (\exists x)B(y, x)) \supset \sim((\exists x)B(y, x) \supset \sim A(y)))$$

Proceeding as in the last example, we obtain

(1) $(y)[(\sim A(y) \supset (\exists x)B(y, x)) \supset \sim((\exists x)B(y, x) \supset \sim A(y))]$
(2) $(y)[(\exists z)(\sim A(y) \supset B(y, z)) \supset \sim((\exists x)B(y, x) \supset \sim A(y))]$
(3) $(y)(x)[(\sim A(y) \supset B(y, x)) \supset \sim((\exists x)B(y, x) \supset \sim A(y))]$
(4) $(y)(x)[(\sim A(y) \supset B(y, x)) \supset \sim(z)(B(y, z) \supset \sim A(y))]$
(5) $(y)(x)[(\sim A(y) \supset B(y, x)) \supset (\exists z) \sim (B(y, z) \supset \sim A(y))]$
(6) $(y)(x)(\exists x_1)[(\sim A(y) \supset B(y, x)) \supset \sim(B(y, x_1) \supset \sim A(y))]$

Again we have arrived at a formula in prenex normal form. Notice that the presence of both $x$ and $x_1$ shows that the original formula is not equivalent to the formula $\sim(\exists y)(\exists x)(\sim A(y) \equiv B(y, x))$: the two parts of the formula might require different values of $x$ and $x_1$ to satisfy them. Another formula equivalent to the given one is

$$(y)[(\exists x)(\sim A(y) \supset B(y, x)) \supset (\exists x_1)(B(y, x_1) \wedge A(y))]$$

One possible interpretation of this is "For all numbers $y$, if there exists a number $x$ such that whenever $y$ is non-positive then $B(y, x)$ holds, then there exists a number $x_1$ such that $y$ is positive and $B(y, x_1)$ holds." Here we have interpreted $A(y)$ as "$y$ is positive," and left the predicate $B$ uninterpreted. From this it should be intuitively clear that the values of $x$ and $x_1$ need not be the same.

As with the normal forms for the propositional calculus, the prenex normal form is not necessarily the shortest equivalent formula, but it does provide a useful standard. In fact, most techniques which have been developed for determining whether or not a formula is valid or satisfiable depend on having a prenex normal form with a certain specified sequence of quantifiers [these are often finite truth table techniques (hence algorithms) for special classes of formulas].

EXERCISES

For each of the following formulas, find an equivalent formula in prenex normal form.

1. $(x)\mathscr{A}(x) \lor \mathscr{B}$
2. $(\exists x)\mathscr{A}(x) \lor \mathscr{B}$
3. $(x)\mathscr{A}(x) \land \mathscr{B}$
4. $(x)\mathscr{A}(x) \equiv \mathscr{B}$
5. $(x)\mathscr{A}(x) \mid \mathscr{B}$
6. $(x)A(x) \supset (\exists y)B(x, y)$
7. $A(x, y) \land (\exists x)B(x) \supset (y)(z)C(x, y, z)$
8. $(x)(\exists y)(A(y, x) \land (y)C(y)) \supset B(x, y)$
9. $(x)(y)(A(z) \supset B(z))$
10. $A(x) \supset (B(y) \supset ((\exists x)C(y) \supset (y)D(x)))$

## 6. AXIOMS AND THEOREMS

The concepts of theorem, proof, and deduction which were developed for the propositional calculus carry over to the first-order predicate calculus quite directly. However, since we do want to work with the quantifiers, certain extensions of these concepts are necessary. We do not wish to venture deeply into this area, but will state the axioms and rules of inference, and comment briefly on the development of this topic.

As we mentioned earlier, we must modify the substitution rules so that we do not by substitution bind variables which should be free. The essence of this is contained in the following definition.

**Definition 6.1.** Let $\mathscr{A}$ be a wff and $t$ a term. We say that $t$ *is free for* $x$ *in* $\mathscr{A}$ if and only if no free occurrences of $x$ in $\mathscr{A}$ lie within the scope of any quantifier on $y$, where $y$ is any of the variables occurring in $t$.

Suppose that we wish to substitute $t$ for $x$ in $\mathscr{A}$. If a free occurrence of $x$ in $\mathscr{A}$ lies within the scope of a quantifier on $y$, where $y$ is some variable of $t$, then by substituting $t$ for $x$ that occurrence of $y$ (which takes the place of a free occurrence of $x$) becomes bound. Thus to say that $t$ is free for $x$

in $\mathscr{A}$ is to say that we may substitute $t$ for all free occurrences of $x$ in $\mathscr{A}$ without danger of binding any previously free variables.

The axioms of a first-order predicate calculus fall into two classes. The first of these are known as the *logical axioms*, and are logically valid formulas. In a *pure* calculus these are the only axioms. An *applied* calculus, or *first-order theory* has in addition a set of *proper axioms*, which deal with the subject matter of the theory. For example, if one were developing an axiomatic set theory, the proper axioms would specify some of the supposed properties of sets and the relations between them. The proper axioms are usually not logically valid, but are valid in some desired interpretation. By specifying the axioms by axiom schema (see Chapter 3, Section 9) we avoid the need for having a substitution rule among the rules of inference. Notice that the first three logical axioms are those of the propositional calculus.

**Axiom 1.** $\mathscr{A} \supset (\mathscr{B} \supset \mathscr{A})$.

**Axiom 2.** $(\mathscr{A} \supset (\mathscr{B} \supset \mathscr{C})) \supset ((\mathscr{A} \supset \mathscr{B}) \supset (\mathscr{A} \supset \mathscr{C}))$.

**Axiom 3.** $(\sim\mathscr{B} \supset \sim\mathscr{A}) \supset ((\sim\mathscr{B} \supset \mathscr{A}) \supset \mathscr{B})$.

**Axiom 4.** $(x)\mathscr{A}(x) \supset \mathscr{A}(t)$, where $t$ is any term free for $x$ in $\mathscr{A}$ $(x)$.

**Axiom 5.** $(x)(\mathscr{A} \supset \mathscr{B}) \supset (\mathscr{A} \supset (x)\mathscr{B})$ if $\mathscr{A}$ is a wff containing no free occurrences of $x$.

In the last two axioms, $x$ is any individual variable. Notice that $x$ is free for $x$ in $(x)\mathscr{A}(x)$, so that Axiom 4 includes all statements of the form $(x)\mathscr{A}(x) \supset \mathscr{A}(x)$.

These are the logical axioms; in a pure theory we have no proper axioms. The necessary rules of inference are two:

1. Modus ponens: From $\mathscr{A}$ and $\mathscr{A} \supset \mathscr{B}$ it is permissible to generate $\mathscr{B}$.

2. Generalization: From $\mathscr{A}$ it is permissible to generate $(x)\mathscr{A}$, where $x$ is any individual variable.

The concepts of theorem and proof, and of deduction, are identical with those of the propositional calculus (Chapter 3, Section 9) except that the axioms, rules of inference, and formulas referred to are those of the first-order predicate calculus.

In the propositional calculus the business of discovering proofs was greatly facilitated by the Deduction Theorem. Unfortunately, this does not carry over completely into the predicate calculus because of the difficulties with substitution.

### Weak Deduction Theorem

If a deduction $\Gamma, \mathscr{A} \vdash \mathscr{B}$ involves no application of the rule of generalization in which the quantified variable is free in $\mathscr{A}$, then $\Gamma \vdash \mathscr{A} \supset \mathscr{B}$.

A somewhat stronger theorem may be stated, but it is more cumbersome, and often this weak version is sufficiently useful.

*Example 6.1.* $\vdash (x)(y)\mathscr{A} \supset (y)(x)\mathscr{A}$.

*Proof:*

| | |
|---|---|
| $(x)(y)\mathscr{A}$ | Hypothesis |
| $(x)(y)\mathscr{A} \supset (y)\mathscr{A}$ | Axiom 4 |
| $(y)\mathscr{A}$ | Modus Ponens |
| $(y)\mathscr{A} \supset \mathscr{A}$ | Axiom 4 |
| $\mathscr{A}$ | Modus Ponens |
| $(x)\mathscr{A}$ | Generalization |
| $(y)(x)\mathscr{A}$ | Generalization |

Thus we have $(x)(y)\mathscr{A} \vdash (y)(x)\mathscr{A}$, and since Generalization was not applied to any variable which was free in $(x)(y)\mathscr{A}$, we may apply the Weak Deduction Theorem to obtain the desired theorem.

The theorem proven in Example 6.1 states that the order in which universal quantifiers appear *together* may be changed at will. The axioms which we have stated make no mention of the existential quantifier, so that this must be introduced by definition: $(\exists x)\mathscr{A}$ being defined to mean $\sim(x) \sim \mathscr{A}$. One may then easily show that the order in which existential quantifiers appear *together* may also be freely changed. However, one cannot generally interchange universal and existential quantifiers: $(x)(\exists y)\mathscr{A}$ is quite a different formula from $(\exists y)(x)\mathscr{A}$.

*Example 6.2.* Let $G(x, y)$ be the predicate "$y$ is greater than $x$," where we are thinking of $x$ and $y$ as integers, and "greater than" in the ordinary sense. Then $(x)(\exists y)G(x, y)$ may be interpreted as "For every integer $x$ there is an integer $y$ which is larger than $x$." This is clearly true. On the other hand, $(\exists y)(x)G(x, y)$ represents the false statement, "There is an integer $y$ which is larger than any integer $x$."

EXERCISES

Prove each of the following theorems of the first-order predicate calculus.

1. $(\exists x)(\exists y)\mathscr{A} \equiv (\exists y)(\exists x)\mathscr{A}$
2. $(x)(\mathscr{A} \supset \mathscr{B}) \supset ((x)\mathscr{A} \supset (x)\mathscr{B})$
3. $(x)(\mathscr{A} \supset \mathscr{B}) \supset (x)(\sim\mathscr{B} \supset \sim\mathscr{A})$
4. $(x)(\mathscr{A} \supset \mathscr{B}) \supset ((\exists x)\mathscr{A} \supset (\exists x)\mathscr{B})$
5. $(x)(\mathscr{A} \wedge \mathscr{B}) \equiv (x)\mathscr{A} \wedge (x)\mathscr{B}$
6. $(\exists x)(\mathscr{A} \supset \mathscr{B}) \equiv (x)\mathscr{A} \supset (\exists x)\mathscr{B}$

**References**

1. Church, A. *Introduction to Mathematical Logic.* Vol. I, Princeton University Press, Princeton, New Jersey, 1956.
2. Mendelson, E. *Introduction to Mathematical Logic.* D. Van Nostrand Company, Princeton, New Jersey, 1964.

3. Rosenbloom, P. C. *The Elements of Mathematical Logic*. Dover Publications, New York, 1950.
4. Stoll, R. R. *Set Theory and Logic*. W. H. Freeman and Company, San Francisco, 1963.
5. Suppes, P. *Introduction to Logic*. D. Van Nostrand Company, Princeton, New Jersey, 1957.

# 7. Formal Languages

## 1. POST LANGUAGES

Virtually all activity associated with the concepts discussed in previous chapters is of one general character: the activity of formally changing strings of symbols into other strings of symbols by certain prescribed rules. One of the first men to recognize this was the American logician Emil Post. In this section we will examine the theory which he developed and which has come to be known as the theory of Post languages; in the following section, we shall look briefly at more recent developments in this area.

We are concerned with operations on a finite non-empty set of symbols called the *alphabet*. These correspond to the set of variables and connectives in the propositional calculus (together with the parentheses), or to the alphabet in ordinary English. There is a finite set of rules which apply to these symbols, and which specify how to combine strings of symbols into *words*, or *sentences* (the wffs of logic, or the words of English). These rules are called *productions*. Generally we also have a set of strings called *axioms* to which we may initially apply the productions. We proceed now with the formal definition of these terms.

**Definition 1.1.** Let $\mathscr{A}$ be a finite set, called the *alphabet*; let $\mathscr{V}$ be the set of all finite strings of symbols in $\mathscr{A}$, called the *vocabulary*. Suppose that the symbols $\alpha_1, \alpha_2, \alpha_3, \ldots,$ are not in $\mathscr{V}$. (Informally, we shall use $\alpha, \beta, \ldots$.) A *string schema* is a finite sequence

$$\alpha_{i_1} s_1 \alpha_{i_2} s_2 \ldots \alpha_{i_{n-1}} s_{n-1} \alpha_{i_n}$$

where each $s_i$ is an element of $\mathscr{V}$, and at least one of the $\alpha$'s actually occurs (one or more may be null, or not occur). A *production* is a sequence $S_1, S_2, \ldots, S_m \to T$, where $m$ is a positive integer, $S_1, S_2, \ldots, S_m$, and $T$ are string schemata, and if $T = \alpha_1 t_1 \alpha_2 t_2 \ldots \alpha_{n-1} t_{n-1} \alpha_n$, then at least one of the $t_i$ is not the null string and each of the $\alpha_i$ which actually occurs in $T$ also occurs in at least one of the $S_j$, $j = 1, \ldots, m$.

The use of the productions is similar to that of the Markov productions, but without the left-to-right orientation. A string schema is thought of as representing the set of all strings which may be obtained by substituting arbitrary strings in $\mathscr{V}$ for each of the $\alpha$'s. Such strings may be said to *satisfy* the schema. If we then have in a given set of strings some strings

144

which satisfy $S_1, \ldots, S_m$, where we have a production $S_1, \ldots, S_m \to T$, we may add to the set the string satisfying $T$ which is obtained by substituting for the $\alpha$'s in $T$ the strings which correspond to these same $\alpha$'s in the $S_1, \ldots, S_m$. (We assume here that if one of the $\alpha$'s occurs in two or more of the $S_j$ it is replaced in all occurrences by the same string.)

*Example 1.1.* Suppose that we have two productions

$$\alpha a \to b a \alpha$$
$$\gamma a \delta b, \, b \alpha a \beta \to \alpha \beta b b \delta,$$

and that the given set of strings consists of the single string $a$. Since this is a string satisfying $\alpha a$ (with $\alpha$ replaced by the null string) we may use the first production to add to the set the string $ba$. This new string satisfies $b \alpha a \beta$, but we have no string satisfying $\gamma a \delta b$, so that we are unable to use the second production yet. However, $ba$ does satisfy $\alpha a$ (with $b$ as the $\alpha$), so that we may use the first production to add to the set another string, namely $bab$. This string no longer satisfies $\alpha a$, but it does satisfy both $\gamma a \delta b$ (with $b$ as $\gamma$, and $\delta$ null) and $b \alpha a \beta$ (with $\alpha$ null and $b$ as $\beta$). Thus we may produce the string $bbb$. We may also use $bab$ to satisfy $\gamma a \delta b$ and $ba$ to satisfy $b \alpha a \beta$, producing the string $bb$. Since there are no further ways to satisfy the string schemata, the final set of words is $\{a, ba, bb, bab, bbb\}$.

Even though the set generated in Example 1.1 is finite, this is not always the case; in fact, in the most interesting situations the generated set has infinitely many members.

**Definition 1.2.** A *Post canonical language* consists of a finite alphabet, a finite set of productions, and a finite subset of the vocabulary, called the set of *axioms*.

It should be remembered that the alphabet itself is a finite subset of the vocabulary, and hence could be used as the set of axioms.

*Example 1.2.* In this example we develop the propositional calculus as a Post canonical language, using the Polish notation. We must have symbols for the propositional variables and the connectives, and various class symbols to distinguish the propositional variables, the well-formed formulas, and the theorems. Since we wish to have an unlimited number of propositional variables while the alphabet is restricted to be finite, we must generate the propositional variables. We take the alphabet to be $\{N, C, A, K, E, p, 1, \vdash, \mathbf{W}, \mathbf{L}\}$. The last three symbols listed will designate respectively the classes of theorems, wffs, and propositional variables (letters). We take just one axiom, $\mathbf{L}p$, and fourteen productions:

1. $\mathbf{L}\alpha \to \mathbf{L}\alpha 1$
2. $\mathbf{L}\alpha \to \mathbf{W}\alpha$
3. $\mathbf{W}a \to \mathbf{W}N\alpha$

4. $W\alpha, W\beta \rightarrow WC\alpha\beta$

5. $WCN\alpha\beta \rightarrow WA\alpha\beta$

6. $WA\alpha\beta \rightarrow WCN\alpha\beta$

7. $WNC\alpha N\beta \rightarrow WK\alpha\beta$

8. $WK\alpha\beta \rightarrow WNC\alpha N\beta$

9. $WNCC\alpha\beta NC\beta a \rightarrow WE\alpha\beta$

10. $WE\alpha\beta \rightarrow WNCC\alpha\beta NC\beta a$

11. $W\alpha, W\beta \rightarrow \vdash C\alpha C\beta\alpha$

12. $W\alpha, W\beta, W\gamma \rightarrow \vdash CC\alpha C\beta\gamma CC\alpha\beta C\alpha\gamma$

13. $W\alpha, W\beta \rightarrow \vdash CCN\alpha N\beta CCN\alpha\beta\alpha$

14. $\vdash\alpha, \vdash C\alpha\beta \rightarrow \vdash\beta$

With the given axiom, the first production specifies that the propositional variables are all strings of the form $p, p1, p11, p111, \ldots$. The second, third, and fourth productions are the rules for forming well-formed formulas, and the fifth through the tenth productions define $A$, $K$, and $E$ in terms of $C$ and $N$. Productions 11, 12, and 13 are the "axioms" of the propositional calculus in the sense of Chapter 3, and production 14 provides us with modus ponens.

**Definition 1.3.** The set of *theorems* of a Post canonical language consists of the axioms of the language together with all strings generated from the axioms by the productions of the language.

**Definition 1.4.** Given a (Post canonical) language $\mathscr{L}$, a language $\mathscr{L}'$ is an *extension* of $\mathscr{L}$ if the alphabet of $\mathscr{L}$ is contained in the alphabet of $\mathscr{L}'$ and each theorem of $\mathscr{L}$ is a theorem of $\mathscr{L}'$.

Thus we may extend a language by adding to it more symbols, productions, and axioms. However, it is not necessary that the extension be done in this way. For example, one can develop a partial propositional calculus by omitting, say, production 13 from the language of Example 1.2. One possible extension of this is then the language of Example 1.2. However, another extension may be obtained by utilizing the language of the example, with productions 11, 12, and 13 replaced by the single production $W\alpha, W\beta, W\gamma, W\delta, W\epsilon \rightarrow \vdash CCCCC\alpha\beta CN\gamma N\delta\gamma\epsilon CC\epsilon\alpha C\delta\alpha$, for it is known that this single expression suffices as an axiom from which the propositional calculus may be developed (see Meredith [4]). Thus an extension of a given language may be derived in quite another way, even to the point that it may be difficult to recognize the fact that it is an extension of the language.

**Definition 1.5.** An extension $\mathscr{L}'$ of a language $\mathscr{L}$ is said to be *conservative* if every theorem of $\mathscr{L}'$ which is a string in the alphabet of $\mathscr{L}$ is also a theorem of $\mathscr{L}$.

Thus a conservative extension of a language is one which does not enlarge the set of theorems which may be stated in the alphabet of the original language: any new theorems involve symbols which are not in the original alphabet.

**Definition 1.6.** A Post canonical language is called *normal* if it has only one axiom and all productions are of the form $s_1\alpha \to \alpha s_2$, where $s_1$ and $s_2$ are specified strings.

Post established the important result that every canonical language has a conservative normal extension [6]. The proof, which is quite long, involves the construction of a series of languages, each of which is shown to be a conservative extension of the previous one, and the last of which is normal. This then provides a standard or normal form for the study of Post languages. As is the case with the logical systems which we have discussed, this is not the only possible standard form. Post, Thue, and others have studied several varieties (see, for example, Davis [3]).

As was suggested at the beginning of this chapter, the various topics which we have discussed are expressible in Post languages. Because of this, one would expect to find unsolvable problems appearing in Post languages, just as the halting problem for Turing machines and other unsolvable problems have appeared in our previous discussions. These take the form of *decision problems* on theorems: to decide whether or not a given expression is a theorem in a given language.

EXERCISES

1. Describe a Post canonical language whose theorems are the individual constants and variables, and the function and predicate letters of the first order predicate calculus, as defined in Chapter 6.

2. Consider the language:
   Alphabet:   $a, b, c, d$
   Axioms:   $a, bc$
   Productions:   $\alpha a \to b a \alpha$
   $\gamma a \delta b, b \alpha a \beta \to \alpha \beta b b \delta$
   $\alpha a \beta b \to \alpha c b \beta$
   $\alpha c \to d a \alpha b$
   $\alpha b \to d c \alpha a$

   Show that the following strings are theorems of this language.
   (a) *dcbba*        (f) *adcdbbbdcda*
   (b) *badcba*       (g) *dcadcbbbdca*
   (c) *badcdab*      (h) *bbbdcdbbb*
   (d) *dcdabbbdcda*  (i) *bbbdcb*
   (e) *adcbbb*       (j) *dabadcbbbdb*

## 2. RECENT ADVANCES IN FORMAL LANGUAGES

With the advent of large digital computers, studies in formal languages were greatly accelerated because of developments in two distinct areas. Attempts to produce successful mechanical translation algorithms showed that the traditional linguistic studies of natural languages inadequately described these languages for this purpose. Thus there was a need to formalize the properties of various natural languages which were pertinent to the construction of such algorithms. On the other hand, the sophisticated programming languages which have been developing along with the computers are themselves formal in nature. As these languages have become more flexible, the translators needed to produce machine language programs from them have become more complex. In order to describe these new languages adequately and to construct translators which will rapidly produce an efficient working program, it has become necessary to study the characteristics of formal languages.

In the description of formal languages, the most important new development has been that of the Backus Normal Form. This is a formalization of a portion of the metalanguage which allows a precise and concise description of the formal language. In the Backus Normal Form, classes or sets of symbol strings are named by using angular brackets, ⟨ ⟩. Whatever appears inside of such brackets is the name for a class of symbol strings. The symbol ::= is used as the definitional symbol: the class whose name appears on the left of this symbol consists of those strings of symbols which are listed on the right. A vertical line ( | ) is used as a separator in definitions, and concatenation in a definitional statement denotes concatenation within the formal language.

*Example 2.1.* The wffs of the propositional calculus may be described with the following Backus Normal Form definitions, assuming only four propositional variables.

$$\langle \text{symbol} \rangle ::= p \mid q \mid r \mid s \mid \supset \mid \sim \mid ( \mid )$$
$$\langle \text{variable} \rangle ::= p \mid q \mid r \mid s$$
$$\langle \text{connective} \rangle ::= \supset \mid \sim$$
$$\langle \text{wff} \rangle ::= \langle \text{variable} \rangle \mid \sim (\langle \text{wff} \rangle) \mid (\langle \text{wff} \rangle) \supset (\langle \text{wff} \rangle)$$

The first of these statements declares that there are eight symbols, which include ( and ). The second and third statements define respectively the classes of variables and connectives. The final statement declares that a wff is any variable; or the symbols ~( followed by any wff, followed by ); or the symbol ( followed by any wff, followed by the symbols ) ⊃ (, followed by any wff, followed by ).

*Example 2.2.* These statements describe the construction of decimal numerals.

$$\langle\text{symbol}\rangle ::= 0 \mid 1 \mid 2 \mid 3 \mid 4 \mid 5 \mid 6 \mid 7 \mid 8 \mid 9 \mid .$$
$$\langle\text{digit}\rangle ::= 0 \mid 1 \mid 2 \mid 3 \mid 4 \mid 5 \mid 6 \mid 7 \mid 8 \mid 9$$
$$\langle\text{integer}\rangle ::= \langle\text{digit}\rangle \mid \langle\text{digit}\rangle\langle\text{integer}\rangle$$
$$\langle\text{decimal numeral}\rangle ::= \langle\text{integer}\rangle \mid \langle\text{integer}\rangle . \mid$$
$$\langle\text{integer}\rangle . \langle\text{integer}\rangle$$

In the third statement, an integer is defined to be a digit, or a digit followed by any previously defined integer. That is, an integer is defined to be a finite sequence of digits. The fourth statement provides for the introduction of the decimal point. Notice that such strings as 0 and 000172 are legitimate integers, so that 0.000172 is a decimal numeral. However, .000172 is not a decimal numeral, according to these definitions.

A good example of the use of the Backus Normal Form to describe a programming language is provided by the ALGOL 60 Report [5].

The direction in which studies of formal languages have progressed has been determined largely by the work of Noam Chomsky and others beginning around 1955–1957. Of key importance in these studies has been the recognition of the importance of phrase structure and context in the comprehension of meaning.

**Definition 2.1.** A *phrase structure grammar* is a finite set of Post productions containing exactly one symbol S which appears only on the left of "→", and a nonempty set of symbols which appear only on the right of "→". A *phrase structure language* is a Post canonical language whose productions constitute a phrase structure grammar, and whose only axiom is the symbol S which appears only on the left side of the productions.

This definition is motivated by the description of the ways in which a sentence may be constructed. The symbol S may be thought of as meaning "sentence," and the productions describe the manners in which sentences may be constructed. The symbols which appear only on the right side of the productions, known as the *terminal symbols*, then become the words of which a sentence is composed.

*Example 2.3.* Alphabet: S, NP, VP, AP, N, V, Adj, Adv, Art, Pred, a, an, ball, colorless, furiously, green, ideas, is, John, old, plays, quietly, sleep, sleeps, square, the.

Axiom:  S

Productions:

| | |
|---|---|
| S → NP VP | AP → Adj N |
| NP → N | VP → V |
| NP → Art AP | VP → V NP |
| NP → Adj AP | VP → V Pred |
| AP → Adj AP | Pred → Adv |

| | |
|---|---|
| N → ball | Adj → green |
| N → ideas | Adj → old |
| N → John | Adj → square |
| V → is | Adv → furiously |
| V → plays | Adv → quietly |
| V → sleep | Art → a |
| V → sleeps | Art → an |
| Adj → colorless | Art → the |

Notice that this language could be more compactly described using the Backus Normal Form. A typical construction ("proof") in this language is the following sequence of symbol strings. S, NP VP, N VP, N V NP, N V N, John V N, John plays N, John plays ball. Other sentences which may be constructed in this language include "John sleeps quietly," and "Colorless old John is square." However, this language is clearly not usable for normal English discourse, since it also permits such sentences as "The square ball plays John," "A idea sleep the ball," and "Colorless green ideas sleep furiously."*

One of the prime problems in devising any mechanical translation algorithm is that of handling contextual references. In the various natural languages the meaning of a particular word often depends on the context in which it occurs; this context may extend even beyond the bounds of the particular sentence in which the occurrence lies. For example, "They are visiting relatives" has two meanings, depending on whether one interprets "visiting" as modifying "relatives" or as a portion of the verb. Yet without knowing the context within which this sentence lies, we cannot tell which is the proper interpretation. This has motivated a study and classification of formal languages which is dependent on various contextual conditions. For example, a *context-free language* is a language in which all productions have the form $s \rightarrow \ldots$, where $s$ is a specified symbol. That is, the productions depend only on specific symbols, and not on the context in which they appear. Thus the language of Example 2.3 is context-free.

We have mentioned that Turing machines, Markov algorithms, and so forth are describable in terms of formal languages. The same holds true for the various types of automata which have been studied. Conversely, it has been shown that for various languages, automata of different types can be constructed which will recognize or accept those and only those strings which are sentences in the language. Moreover, these automata have a natural classification which parallels that of the formal languages.

---

* This last sentence is due to Chomsky in [1].

Exercises

1. Describe a Post canonical language whose theorems are just those strings described by the Backus Normal Form statements of Example 2.2.

2. Describe the language of Example 2.3 using Backus Normal Form statements.

## References

1. Chomsky, N. "Three models for the description of language." *Proceedings of the 1956 Symposium on Information Theory, IRE Transactions on Information Theory IT-2*, **3** (Sept. 1956), 113–124.

2. Chomsky, N. *Syntactic Structures*. Mouton and Company, The Hague, Netherlands, 1957.

3. Davis, Martin. *Computability and Unsolvability*. McGraw-Hill Book Company, New York, 1958.

4. Meredity, C. A. "Single axioms for systems (C, N), (C, O), and (A, N) of the two-values propositional calculus." *Journal of Computing Systems* **1**, No. 3 (1953), 155–164.

5. Naur, P., ed. "Report on the algorithmic language ALGOL 60." *Communications of the Association for Computing Machinery*, **3** (1960), 299–314.

6. Post, E. L. "Formal reductions of the general combinatorial decision problem." *American Journal of Mathematics* **65** (1943), 197–215.

7. Rosenbloom, P. C. *The Elements of Mathematical Logic*. Dover Publications, New York, 1950.

8. "Proceedings of a working conference on mechanical language structures." *Communications of the Association for Computing Machinery* **7** (1964), 51–136.

9. "Structure of language and its mathematical aspects." *Proceedings of Symposia in Applied Mathematics*, Vol. XII. American Mathematical Society, Providence, Rhode Island, 1961.

# 8. A Brief History

In this book we have examined some of the basic ideas behind the manner in which people today, with the devices at their disposal, solve problems. The setting is mathematical because of two factors—that people have understood quite well the problem-solving process when the problems were those of classical mathematics, and that modern mathematics is expanding its applications into areas of the sciences and humanities which have traditionally been considered nonmathematical and qualitative in nature. There has been continuing commentary in both the popular and the scholarly press about the impact of computers on our society, covering the gamut from automation to artificial intelligence. Thus it is useful to look briefly at the history of developments in this area in order that we may better appreciate the situation today.

If we call our subject "logic, algorithms, and languages," we see that man's concern with this area is ancient, although this concern has not always been mathematical. The need to translate languages first arose when two social groups with different languages came in contact. Yet until very recently language translation and linguistic studies in general have proceeded on a thoroughly nonmathematical course. The importance of having algorithms or methods of calculating was known to the ancient Egyptians and Greeks. The name "algorithm" itself is derived from al-Khowarazmi, the name of an Arab mathematician (A.D. 780–c.850) who wrote a treatise on algebra about A.D. 830. The use of the term to denote finite calculations in general (as opposed to calculations with Arabic numerals), as we have used it, is again a recent development.

The evolution of logic began with Aristotle, 384–322 B.C., and other ancient Greeks. As in geometry and other fields, the Grecian ideas of logic completely dominated the field for the next 2000 years. Then around 1666 Gottfried Wilhelm von Leibniz (1646–1716) proposed what he called in various works the "calculus ratiocinator," or "logica mathematica," or "logistica." Although Leibniz never developed these ideas to any great extent, his writings contain the germ of much of mathematical logic.

Perhaps significantly, the art or science of computing moved forward at the same time, and then like logic, lay largely dormant until the nineteenth century. In 1642 Blaise Pascal (1623–1662) constructed the first mechanical adding machine; Leibniz in 1671 built a device which could also multiply. Incidentally, Leibniz also introduced into mathematical terminology the word "function," and was the first to recognize the importance of the binary numeral system (aside from possibly the ancient Chinese).

The next important event in our history occurred in 1812, when the English inventor Charles Babbage (1792–1871) designed a "difference engine" whose purpose was to aid in the construction of mathematical tables. The British government supported the project, but then abandoned it after several years of unsuccessful attempts to build the device. In 1833 Babbage proposed an "analytical engine" which had mechanical storage of data and instructions analogous to the electronic storage of today's computers. The failure of this device, as of the difference engine, was due not to errors in conception, but rather to the fact that machine tools and technology of the day were not adequate to provide the precision components required.

Although the field of computing lay quiet for another century, work in logic progressed rapidly. In 1847 Augustus De Morgan (1806–1871) published a treatise, *Formal Logic*. In the following year George Boole (1815–1864) wrote *The Mathematical Analysis of Logic*, following this in 1854 with *An Investigation of the Laws of Thought*. Although none of these works contained all of the basic elements of logic as we understand it today (Boole, for example, considered only the exclusive use of "or" in developing his algebraic system), within the next forty to fifty years a number of excellent logicians expanded this base to essentially its present form. This period in the development of logic culminated in 1910–1913 with the publication of *Principia mathematica* by Alfred North Whitehead (1861–1947) and Bertrand Russell (1872–    ). This was a monumental attempt to complete a program which Leibniz had suggested, that of providing a logical basis for all of mathematics. The work includes set theory as it had been developed by Georg Cantor (1845–1918) and others, and some portions of geometry, but was never carried further.

By the time of Whitehead and Russell, mathematical logic had reached a healthy adolescence and was growing rapidly into maturity. Since that time the study of logic has developed in several different directions, the one in which we are interested bringing us to algorithms and computers. During the latter half of the nineteenth century, mathematicians had been exploiting and extending the "axiomatic method"—that of stating axioms and then deriving theorems formally from these. Although much of this work had been done in set theory, geometry, and logic, by the turn of the century it seemed entirely reasonable to many mathematicians that eventually all of the various parts of mathematics could be axiomatized. The extreme viewpoint on this matter was that not only could all of the parts be axiomatized, but that mathematics as a whole could actually be axiomatized, in a consistent, relatively simple manner. The German mathematician David Hilbert (1862–1943) held views along these lines. In 1900 Hilbert proposed a list of 23 problems whose solution he regarded as essential to the advancement of mathematics [7]. Many of these problems have since been solved; many have led to other profound problems; some are still unsolved. A few of the problems have been shown to depend strongly on logical concepts, and thus have influenced the development of logic.

The program which Hilbert had in mind was the development of a logico-mathematical system within which all of mathematics would be embedded, and which was provably consistent. That is, it would be possible to prove that from the axioms of this system we could never prove a contradiction as a theorem. In 1931 this program received a severe and probably mortal blow. A young Austrian logician, Kurt Gödel (1906– ) published a paper containing the following result [5]. Any formal system which is sufficiently general to completely contain elementary number theory (a portion of mathematics dealing with the arithmetic of the integers) must either contain formulas which are neither provable nor disprovable within the system or else be inconsistent.

To restate this, Gödel showed that no matter how strong a formal system one invented, there would remain problems which could be formulated within the system, but which were unsolvable in the system. That is, there are mathematical problems which are bound to remain unsolved—not merely because man has not discovered how to solve them, but because man is apparently inherently incapable of discovering how to solve them. As Post has written, "Like the classic unsolvability proofs, these proofs are of unsolvability by means of given instruments. What is new is that ... these instruments, in effect, seem to be the only instruments at man's disposal." [11]

Thus while Gödel's work destroyed Hilbert's program, it led to the rapid development of a neglected field of mathematical endeavor—that of determining what are valid methods for solving problems. The basic questions here are exactly what we mean when we say that we have an algorithm for solving a problem, that we can "effectively compute" a function, or that we can "effectively enumerate" a set.

In the 1930's a number of excellent mathematicians and logicians addressed themselves to these questions, with the result that several different concepts were proposed as satisfactory answers to them. Among the more important of these concepts are those of recursive functions, introduced by Gödel [5, 6], λ-definability, due to Church [2], and computability as postulated by Turing [12] and independently by Post [10].

While these concepts are superficially quite different, the fact that they are equivalent was suspected by the people involved, and shown originally in two papers by Kleene [9] and Turing [13].

After this initial spurt of activity, work in the area of computability became gradually more involved with the finer points of the theory. Meanwhile the world of computers was slowly coming into being. While analog computers had been discussed and constructed several years earlier, and the punched card had been invented by Herman Hollerith as long ago as 1886, digital computers did not appear until 1940. In that year Bell Telephone Laboratories' Complex Computer was put into operation. It was a special-purpose machine, designed to handle the calculations involving complex numbers which electrical engineers encounter in their work. The first general-purpose digital computer, the Harvard Mark I, was put into operation four years later, followed by the ENIAC in 1946, completed at the Moore School of Electrical Engineering of the University of Pennsylvania. The first large computer followed this in 1952, although IBM had several smaller computers available earlier.

Since 1952, computers both large and small have appeared profusely throughout the United States and the rest of the world. This fact has given direction and impetus to studies in linguistics, algorithm theory, and logic, and will undoubtedly continue to do so for many years. The work in linguistics has been mentioned in the previous chapter. In algorithm theory, the work of Markov is quite computer-oriented, and Hao Wang has suggested what is essentially a programmed version of a Turing machine [14]. Moreover, the development of the theory of automata has often been motivated by a desire to construct mathematical models of digital computers, with recent works in this area taking into account more and more of the structural characteristics of these machines.

In addition to the use of Boolean algebra in the design of computers, and logical concepts in the development of algorithms, logic itself has entered into two developments in computing. Although there is no algorithm for determining whether or not a given formula of the first-order predicate calculus is a theorem, for limited classes of such formulas, algorithms do exist. Recently computers have been used in the development of these algorithms, and in the development of efficient semi-algorithms for other classes of formulas. Probably of more significance in the evolution of computers is the area of theorem proving. For this the computer is usually supplied with an axiomatization of either the propositional calculus or of geometry, some rules of inference, and some more-or-less intuitive guidelines, along with a suitable program. With this background, the computer is then supposed to prove theorems. Attempts at this have been fairly successful, but not spectacular. Nevertheless, these attempts have been combined into one of the wellsprings of the area of artificial intelligence—the effort to make a computer "think" like a person.

The evolution of computers has been so rapid that it is foolish to hazard an opinion on their development within the next ten or twenty years. However, the problems of communicating with the computer and with other people (in part through a computer) will remain with us. If these and other problems are to be solved with the aid of a computer, it appears that the only basic tool which we have at our disposal is the algorithm in one of its various forms.

## References

1. Boole, G. *An Investigation of the Laws of Thought.* Dover Publications, New York, 1958.
2. Church, A. "An unsolvable problem of elementary number theory." *The American Journal of Mathematics* **58** (1936), 345–363.
3. Church, A. "The calculi of lambda-conversion." *Annals of Mathematics Studies, No.* 6, Princeton University Press, Princeton, New Jersey, 1951.
4. Davis, Martin, ed. *The Undecidable.* Raven Press, Hewlett, New York, 1965. (This book contains reprints of [2], [5] (in translation), [10], [11], [12], and other basic papers.)
5. Gödel, K. "Über formal unentscheidbare Sätze der principia Mathematica und verwandter Systeme, I." *Monatschefte für Mathematik und Physik* **38** (1931), 173–198.
6. Gödel, K. *On Undecidable Propositions of Formal Mathematical Systems.* Princeton University Press, Princeton, New Jersey, 1934.
7. Hilbert, D. "Mathematical problems." *Bulletin of the American Mathematical Society* **8** (1901–02), 437–479.
8. Hilbert, D. and P. Bernays. *Grundlagen der Mathematik.* Vol. I (1934), Vol. II (1939). Springer-Verlag, Berlin. (Reprinted by Edwards Bros., Ann Arbor, Michigan, 1944.)
9. Kleene, S. "λ-definability and recursiveness." *Duke Mathematics Journal* **2** (1936), 340–353.
10. Post, E. L. "Finite combinatory processes, formulation I." *The Journal of Symbolic Logic* **1** (1936), 103–105.
11. Post, E. L. "Recursively enumerable sets of positive integers and their decision problems." *Bulletin of the American Mathematical Society* **50** (1944), 284–316.
12. Turing, A. M. "On computable numbers, with an application to the Entscheidungsproblem." *Proceedings of the London Mathematical Society* (2) **42** (1936–7), 230–265.
13. Turing, A. N. "Computability and λ-definability." *The Journal of Symbolic Logic* **2** (1937), 153–163.
14. Wang, Hao. "A variant to Turing's theory of calculating machines." *Journal of the Association for Computing Machinery* **4** (1957), 63–92.
15. Whitehead, A. N., and B. Russell. *Principia mathematica,* 3 vols., Cambridge, England, 1910–1913: second edition 1925–1927.

# Answers to the Exercises

## Chapter 1, Section 1

1. (a) The set consisting of the first five letters of the English alphabet.

   (b) The set of numbers $p \cdot q$, where $p$ and $q$ each have any of the values 1, 5, or 10.

   (c) The set of integral multiples of three.

   (d) The set consisting of the planets Jupiter, Saturn, and Uranus.

2. (a) {a,c,e,i,l,n,o,t,u,w}.

   (b) This answer depends on the year. For the year 1967 the answer is {January, October}; for 1968 the answer is {September, December}.

   (c) $\{x \mid x$ is a real number and $x \neq 0\}$.

   (d) $\{\varnothing, \{(), ()\}, \{(,)\}\}$, where $\varnothing$ is the "empty" set discussed in the next section.

## Chapter 1, Section 2

1. (a) $A$ consists of all real numbers less than 4, together with the numbers 5 and 6.

   (b) $A$ consists of all rational numbers less than 4, together with the numbers 5 and 6.

   (c) $A$ consists of all integers less than 4, together with the numbers 5 and 6.

   (d) $A = \{1, 2, 3, 5, 6\}$ (include 0 if you consider this positive).

   (e) $A = U$.

   (f) $A$ consists of all odd integers less than or equal to 5.

   (g) $A$ consists of all even integers less than or equal to 6, except 4.

   (h) $A = \varnothing$.

2. (a) {0, 3, 5, 8, 9}.

   (b) {0, 2, 3, 5, 6, 10}.

   (c) {1, 2, 3, 4, 5, 6, 7, 8, 9, 10}.

   (d) $\varnothing$.

   (e) $U$.

3. (a) $\{i, o\}$.

   (b) $\{a, e\}$.

   (c) $\varnothing$.

   (d) $U$.

4. (a) $\{4, 7, 10\}$.

   (b) $\{2, 5\}$.

   (c) $\varnothing$.

   (d) $\{a, 17\}$.

   (e) $A$.

   (f) $\varnothing$.

   (g) $\varnothing$.

5. (a) $A \subseteq B$.

   (b) $B \subseteq A$.

   (c) $A = B$, hence each is a subset of the other.

   (d) $A \subseteq B$.

   (e) $A = B$, hence each is a subset of the other.

   (f) Neither is a subset of the other.

## Chapter 1, Section 3

1. (a) $\{1, 2, 5, 8, 10\}$.

   (b) $B$.

   (c) $\{a, *, \$, 1, 2, 3\}$.

   (d) $B$.

   (e) $\{1, 2, 3, 4, 5, 6\}$.

2. (a) $\{1, 8, 10\}$.

   (b) $\{7\}$.

   (c) $\{a, *, \$, 1, 2, 3\}$.

   (d) $B$.

   (e) $\{1, 2, 3, 4, 5, 6\}$.

3. (a) $\{2, 5\}$.

   (b) $A$.

   (c) $\varnothing$.

   (d) $\varnothing$.

   (e) $\varnothing$.

4. Refer to Figures 3.1 and 3.2.

   (a) $B \cap C$:  regions 4, 7

   $A \cap (B \cap C)$:  region 4

   $A \cap B$:  regions 3, 4

   $(A \cap B) \cap C$:  region 4

   Therefore equality.

(b) $B \cap C$:   regions 4, 7
$A \cup (B \cap C)$:   regions 2, 3, 4, 5, 7
$A \cup B$:   regions 2, 3, 4, 5, 6, 7
$A \cup C$:   regions 2, 3, 4, 5, 7, 8
$(A \cup B) \cap (A \cup C)$:   regions 2, 3, 4, 5, 7
Therefore equality.

(c) $(B \cap C)$:   regions 4, 7
$A \triangle (B \cap C)$:   regions 2, 3, 5, 7
$A \triangle B$:   regions 2, 5, 6, 7
$A \triangle C$:   regions 2, 3, 7, 8
$(A \triangle B) \cap (A \triangle C)$:   regions 2, 7
Therefore equality does not hold in general.

5. $A \cap B = A$
$A - B = \varnothing$
$A \triangle B = \overline{A} \triangle \overline{B}$, or $A \triangle \overline{B} = \overline{A} \triangle B$, or $A \triangle B = B - A$

## Chapter 1, Section 4

1. $A \cup B = (A \triangle B) \triangle (A \cap B)$
$A - B = A \triangle (A \cap B)$
$\overline{A} = U \triangle A.$

2. $A \cap B = \overline{\overline{A} \cup \overline{B}}$
$A \triangle B = \overline{\overline{A} \cup B} \cup \overline{A \cup \overline{B}}$
$A - B = \overline{\overline{A} \cup B}.$

3. (a) $\varnothing, \{0\}, \{a\}, \{\#\}, \{2\}, \{0, a\}, \{0, \#\}, \{0, 2\}, \{a, \#\}, \{a, 2\}, \{\#, 2\}, \{0, a, \#\},$
$\{0, a, 2\}, \{0, \#, 2\}, \{a, \#, 2\}, A.$

(b) $2^{10}$, or 1024.

4. (a)

| $\triangle$ | $\varnothing$ | $\{T\}$ | $\{F\}$ | $\{T,F\}$ |
|---|---|---|---|---|
| $\varnothing$ | $\varnothing$ | $\{T\}$ | $\{F\}$ | $\{T,F\}$ |
| $\{T\}$ | $\{T\}$ | $\varnothing$ | $\{T,F\}$ | $\{F\}$ |
| $\{F\}$ | $\{F\}$ | $\{T,F\}$ | $\varnothing$ | $\{T\}$ |
| $\{T,F\}$ | $\{T,F\}$ | $\{F\}$ | $\{T\}$ | $\varnothing$ |

| $\cap$ | $\varnothing$ | $\{T\}$ | $\{F\}$ | $\{T,F\}$ |
|---|---|---|---|---|
| $\varnothing$ | $\varnothing$ | $\varnothing$ | $\varnothing$ | $\varnothing$ |
| $\{T\}$ | $\varnothing$ | $\{T\}$ | $\varnothing$ | $\{T\}$ |
| $\{F\}$ | $\varnothing$ | $\varnothing$ | $\{F\}$ | $\{F\}$ |
| $\{T,F\}$ | $\varnothing$ | $\{T\}$ | $\{F\}$ | $\{T,F\}$ |

(b) Let $A = \{0\}$       $F = \{0,\#\}$       $K = \{0,a,\#\}$
$B = \{a\}$       $G = \{0,2\}$       $L = \{0,a,2\}$
$C = \{\#\}$       $H = \{a,\#\}$       $M = \{0,\#,2\}$
$D = \{2\}$       $I = \{a,2\}$       $N = \{a,\#,2\}$
$E = \{0,a\}$       $J = \{\#,2\}$

| △ | ∅ | A | B | C | D | E | F | G | H | I | J | K | L | M | N | U₂ |
|---|---|---|---|---|---|---|---|---|---|---|---|---|---|---|---|---|
| ∅ | ∅ | A | B | C | D | E | F | G | H | I | J | K | L | M | N | U₂ |
| A | A | ∅ | E | F | G | B | C | D | K | L | M | H | I | J | U₂ | N |
| B | B | E | ∅ | H | I | A | K | L | C | D | N | F | G | U₂ | J | M |
| C | C | F | H | ∅ | J | K | A | M | B | N | D | E | U₂ | G | I | L |
| D | D | G | I | J | ∅ | L | M | A | N | B | C | U₂ | E | F | H | K |
| E | E | B | A | K | L | ∅ | H | I | F | G | U₂ | C | D | N | M | J |
| F | F | C | K | A | M | H | ∅ | J | E | U₂ | G | B | N | D | L | I |
| G | G | D | L | M | A | I | J | ∅ | U₂ | E | F | N | B | C | K | H |
| H | H | K | C | B | N | F | E | U₂ | ∅ | J | I | A | M | L | D | G |
| I | I | L | D | N | B | G | U₂ | E | J | ∅ | H | M | A | K | C | F |
| J | J | M | N | D | C | U₂ | G | F | I | H | ∅ | L | K | A | B | E |
| K | K | H | F | E | U₂ | C | B | N | A | M | L | ∅ | J | I | G | D |
| L | L | I | G | U₂ | E | D | N | B | M | A | K | J | ∅ | H | F | C |
| M | M | J | U₂ | G | F | N | D | C | L | K | A | I | H | ∅ | E | B |
| N | N | U₂ | J | I | H | M | L | K | D | C | B | G | F | E | ∅ | A |
| U₂ | U₂ | N | M | L | K | J | I | H | G | F | E | D | C | B | A | ∅ |

| ∩ | ∅ | A | B | C | D | E | F | G | H | I | J | K | L | M | N | U₂ |
|---|---|---|---|---|---|---|---|---|---|---|---|---|---|---|---|---|
| ∅ | ∅ | ∅ | ∅ | ∅ | ∅ | ∅ | ∅ | ∅ | ∅ | ∅ | ∅ | ∅ | ∅ | ∅ | ∅ | ∅ |
| A | ∅ | A | ∅ | ∅ | ∅ | A | A | A | ∅ | ∅ | ∅ | A | A | A | ∅ | A |
| B | ∅ | ∅ | B | ∅ | ∅ | B | ∅ | ∅ | B | B | ∅ | B | B | ∅ | B | B |
| C | ∅ | ∅ | ∅ | C | ∅ | ∅ | C | ∅ | C | ∅ | C | C | ∅ | C | C | C |
| D | ∅ | ∅ | ∅ | ∅ | D | ∅ | ∅ | D | ∅ | D | D | ∅ | D | D | D | D |
| E | ∅ | A | B | ∅ | ∅ | E | A | A | B | B | ∅ | E | E | A | B | E |
| F | ∅ | A | ∅ | C | ∅ | A | F | A | C | ∅ | C | F | A | F | C | F |
| G | ∅ | A | ∅ | ∅ | D | A | A | G | ∅ | D | D | A | G | G | D | G |
| H | ∅ | ∅ | B | C | ∅ | B | C | ∅ | H | B | C | H | B | C | H | H |
| I | ∅ | ∅ | B | ∅ | D | B | ∅ | D | B | I | D | B | I | D | I | I |
| J | ∅ | ∅ | ∅ | C | D | ∅ | C | D | C | D | J | C | D | J | J | J |
| K | ∅ | A | B | C | ∅ | E | F | A | H | B | C | K | E | F | H | K |
| L | ∅ | A | B | ∅ | D | E | A | G | B | I | D | E | L | G | I | L |
| M | ∅ | A | ∅ | C | D | A | F | G | C | D | J | F | G | M | J | M |
| N | ∅ | ∅ | B | C | D | B | C | D | H | I | J | H | I | J | N | N |
| U₂ | ∅ | A | B | C | D | E | F | G | H | I | J | K | L | M | N | U₂ |

## Chapter 1, Section 5

1. $A ✗ B = \{\langle 1,2\rangle,\langle 1,4\rangle,\langle 3,2\rangle,\langle 3,4\rangle,\langle 5,2\rangle,\langle 5,4\rangle\}.$
$B ✗ A = \{\langle 2,1\rangle,\langle 2,3\rangle,\langle 2,5\rangle,\langle 4,1\rangle,\langle 4,3\rangle,\langle 4,5\rangle\}.$

$A \times A = \{\langle 1,1\rangle,\langle 1,3\rangle,\langle 1,5\rangle,\langle 3,1\rangle,\langle 3,3\rangle,\langle 3,5\rangle,\langle 5,1\rangle,\langle 5,3\rangle,\langle 5,5\rangle\}$.
$B \times B = \{\langle 2,2\rangle,\langle 2,4\rangle,\langle 4,2\rangle,\langle 4,4\rangle\}$.

2. $P = \{\langle a,b\rangle,\langle a,i\rangle,\langle a,k\rangle,\langle a,t\rangle,\langle a,z\rangle,\langle b,i\rangle,\langle b,k\rangle,\langle b,t\rangle,\langle b,z\rangle,\langle i,k\rangle,\langle i,t\rangle,$
$\langle i,z\rangle,\langle k,t\rangle,\langle k,z\rangle,\langle t,z\rangle\}$.

3. $E = \{\langle \text{the},1,2\rangle,\langle \text{quick},2,3\rangle,\langle \text{brown},1,4\rangle,\langle \text{fox},1,2\rangle,\langle \text{jumps},1,4\rangle,$
$\langle \text{over},2,2\rangle,\langle \text{lazy},2,2\rangle,\langle \text{dog},1,2\rangle\}$.

4. (a) $\{\langle \text{son,father}\rangle,\langle \text{mother,daughter}\rangle,\langle \text{brother,father}\rangle,\langle \text{mother},1\rangle,$
$\langle 11,2\rangle,\langle \text{brother},1\rangle,\langle 2,\text{father}\rangle\}$.
The relation is not a mapping.
   (b) $\{\langle 1,a\rangle,\langle 1,b\rangle,\langle 2,c\rangle,\langle 3,d\rangle\}$.
The relation is a mapping, but not one-to-one.
   (c) $\{\langle 1,a\rangle,\langle 2,b\rangle,\langle 3,b\rangle,\langle 4,c\rangle\}$.
The relation is not a mapping, although its inverse is.
   (d) $\{\langle 1,a\rangle,\langle 2,b\rangle,\langle 3,c\rangle,\langle 4,d\rangle\}$.
The relation is a one-to-one mapping.

5. (a) Yes.
   (b) No.
   (c) Yes.

## Chapter 1, Section 6

1.

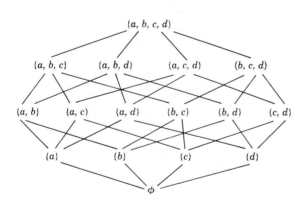

## Chapter 2, Section 1

1. Let $a \# b = (a' \cdot b) + (a \cdot b')$, and let $(-a)$ be $a$.
   Then $a \# (-a) = (a' \cdot a) + (a \cdot a')$
$$= 0 + 0$$
$$= 0$$

2.

| $\alpha \# \beta$ | 0 | $a$ | $b$ | 1 |
|---|---|---|---|---|
| 0 | 0 | $a$ | $b$ | 1 |
| $a$ | $a$ | 0 | 1 | $b$ |
| $b$ | $b$ | 1 | 0 | $a$ |
| 1 | 1 | $b$ | $a$ | 0 |

| $\alpha + \beta$ | 0 | $a$ | $b$ | 1 |
|---|---|---|---|---|
| 0 | 0 | $a$ | $b$ | 1 |
| $a$ | $a$ | $a$ | 1 | 1 |
| $b$ | $b$ | 1 | $b$ | 1 |
| 1 | 1 | 1 | 1 | 1 |

| $\alpha \cdot \beta$ | 0 | $a$ | $b$ | 1 |
|---|---|---|---|---|
| 0 | 0 | 0 | 0 | 0 |
| $a$ | 0 | $a$ | 0 | $a$ |
| $b$ | 0 | 0 | $b$ | $b$ |
| 1 | 0 | $a$ | $b$ | 1 |

## Chapter 2, Section 2

1. $a + a'b = (a + ab) + a'b$
$= a + (ab + a'b)$
$= a + (a + a')b$
$= a + 1 \cdot b$
$= a + b$

2. Since $a'(a')' = 0$, $a' + (a')' = 1$, $a'a = aa' = 0$, and $a' + a = a + a' = 1$, it follows that $(a')' = a$, by the uniqueness postulated in C1.

3. Suppose $a \leq b$. Then $a + b = b$.
$$0 = bb' = (a + b)b'$$
$$= ab' + bb'$$
$$= ab' + 0$$
$$= ab'.$$

Suppose $ab' = 0$. Then
$$a + b = a \cdot 1 + b$$
$$= a(b + b') + b$$
$$= ab + ab' + b$$
$$= ab + 0 + b$$
$$= b + ab + 0$$
$$= b + ab$$
$$= b.$$

Hence $a \leq b$.

4. Suppose $a \leq b$. Then $a + b = b$.
$$b + a' = a + b + a'$$
$$= b + a + a'$$
$$= b + 1$$
$$= 1.$$

Suppose $b + a' = 1$. Then

$$\begin{aligned} a &= a \cdot 1 \\ &= a(b + a') \\ &= ab + aa' \\ &= ab + 0 \\ &= ab. \end{aligned}$$

Hence $b = b + ab = b + a = a + b$, or $a \le b$.

5. Suppose there are three distinct elements in a Boolean algebra, namely, 0, 1, and $a$. (We know that 0 and 1 are both present and are distinct, by Theorem 2.6.) Consider $a'$. From the equations $a + a' = 1$ and $aa' = 0$ it follows that $a'$ cannot be any of the elements 0, $a$, or 1. Hence a fourth element must exist.

6. From the equations $a + a' = 1$ and $aa' = 0$ it follows that for each $a$, $a'$ is distinct from $a$. From these same equations, if $a \ne b$ then $a' \ne b'$. For if $a' = b' = c$, then $c' = (a')'$ and $c' = (b')'$, or $a = (a')' = (b')' = b$. Thus the elements of any Boolean algebra with more than one element must occur in distinct pairs $(0,1)$, $(a,a')$, $(b,b')$, . . . .

## Chapter 2, Section 3

1. $f(x) = ax + a'x'$.
2. $f(x,y,z) = f(1,1,1)xyz + f(1,1,0)xyz' + f(1,0,1)xy'z$
   $\qquad\qquad + f(1,0,0)xy'z' + f(0,1,1)x'yz + f(0,1,0)x'yz'$
   $\qquad\qquad + f(0,0,1)x'y'z + f(0,0,0)x'y'z'$.
3. (a) $f(x) = 0x + 0x'$.
   (b) $f(x,y) = bxy + 1 \cdot xy' + (a + b)x'y + ax'y'$.
   (c) $f(x,y,z) = 1 \cdot xyz + 1 \cdot xyz' + 1 \cdot xy'z + axy'z'$
   $\qquad\qquad + 1 \cdot x'yz + 0 \cdot x'yz' + 1 \cdot x'y'z + 1 \cdot x'y'z'$.
4. $f(a',c,b) = a'bc + a'bc' + abc + abc' + ab'c'$
   $\qquad\qquad = ac' + b$.

## Chapter 2, Section 4

1. (a) $x + y'$.
   (b) $x + y + z'$.
   (c) $wx + w'y + x'z$.
   (d) $wxz + x'yz + w'z' + xy'$ (The first two terms could be replaced by $wyz + x'yz$, or by $wyz + w'x'y$.)
2. (a) $wx + w'y + x'z$.
   (b) $wxz + x'yz + w'z' + xy'$.
   (c) $vwx + vwy + v'wx'z + v'x'yz$.
   (d) $vxz + vx'y' + v'x'z' + wxyz + wx'z'$.

## Chapter 2, Section 5

1. (a)

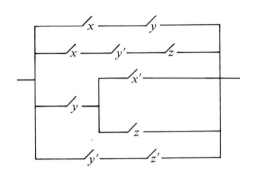

(b)

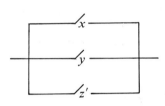

2. (a) $(x + x'y)z + y(xz' + z + x'(y' + z))$.

(b) $(x'(y + y'z) + xyz')w + (xyw' + zwx')y + x'yz'w'$.

(The last term represents a sneak path which will not be present if the flow through the gate marked $z'$ can only be from left to right.)

3.

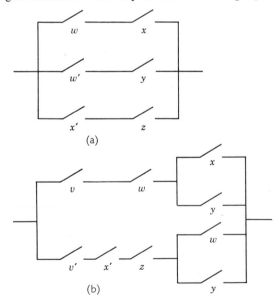

## Chapter 3, Section 1

1. (a) $p$:   The cat is blue.
    $q$:   The cat eats green cheese.
    $p \wedge q$.
  (b) $p$:   Tomorrow is Tuesday.
    $q$:   Tomorrow is Wednesday.
    $p \vee q$, or $p \not\equiv q$.
  (c) $p$:   The Yankees will win the pennant.
    $p \vee {\sim}p$, or $p \not\equiv {\sim}p$.
  (d) $p$:   It is raining.
    $q$:   I will stay home.
    $p \supset q$.
  (e) $p$:   It is raining.
    $q$:   I stay home.
    $p \supset q$.
  (f) $p$:   Tomorrow is Tuesday.
    $q$:   It rains.
    $r$:   It rains on Wednesday.
    $s$:   The garden will dry up.
    $(p \wedge {\sim}q) \supset (r \vee s)$.
  (g) $p$:   The book should be about chemistry.
    $q$:   It should be about biology.
    $r$:   It should be about fungi.
    $s$:   It should be about bacteria.
    $(p \vee q) \wedge (q \supset (r \vee s))$.

## Chapter 3, Section 2

1. (a)  A wff.
  (b)  Not a wff.
  (c)  Not a wff.
  (d)  Not a wff.
  (e)  Not a wff.
  (f)  A wff.
2. (a) $((p) \equiv (q)) \equiv (r)$.
  (b) $(({\sim}(p)) \vee (r)) \equiv (((p) \supset ((q) \wedge ({\sim}(q)))) \supset (r))$.
  (c) $(({\sim}(p)) \vee (r)) \equiv (((p) \supset (q)) \wedge (({\sim}(q)) \supset (r)))$.
  (d) $(({\sim}(((p) \wedge (q)) \vee (r))) \supset (p)) \equiv (q)$.
  (e) $(((p) \vee (r)) \supset ((p) \vee (q))) \supset ((q) \vee (r))$.

### Chapter 3, Section 3

1. $p$ $\quad$ $p \supset (p \supset p) \supset \cdot p$

| | | |
|---|---|---|
| T | | T |
| F | T | F F |

Neither tautology nor contradiction.

2. $q$ $\quad q \quad (p \wedge \sim q) \supset (\sim \sim p \equiv q)$

| T | T | F | T | |
|---|---|---|---|---|
| T | F | T | F | F |
| F | T | F | T | |
| F | F | F | T | |

Neither tautology nor contradiction.

3. Neither:  four lines T, four F.
4. Neither:  six lines T, two F.
5. Neither:  four lines T, four F.
6. Neither:  five lines T, three F.
7. Tautology.
8. Neither:  six lines T, two F.
9. Neither:  seven lines T, one F.
10. Neither:  four lines T, four F.
11. Neither:  five lines T, three F.
12. Neither:  six lines T, two F.
13. Neither:  two lines T, six F.
14. Neither:  three lines T, thirteen F.
15. Neither:  seven lines T, nine F.

### Chapter 3, Section 4

1. V:  The painting is of value.
   S:  The painting is at least 600 years old.
   H:  It was painted with handground pigments.
   L:  It was painted silk or on linen.
   $V \supset \cdot S \vee (\sim S \wedge H) \vee ((S \vee \sim S) \wedge L)$.
This statement is true unless V is true, and S, H, and L are all false.

2. W:  We shall go to the meeting.
   F:  Funds are available.
   C:  The meeting interferes with work.
   S:  There is a related session.
   $W \equiv F \wedge (\sim C \vee S)$.

This statement is true only under the following conditions:
- (i) W and F both true, and either C false or S true.
- (ii) W and F both false.
- (iii) W and S false, F and C true.

3. D: A person can drown in three inches of water.
   U: A person is unconscious.
   L: A person is lying face down in the water.
   M: A person's mustache is on fire.
   T: A person is trying to extinguish the blaze.

This sentence demonstrates the ambiguity of the English language. If "provided that" is regarded as specifying a necessary condition, the symbolism is

$$D \supset ((U \wedge L) \vee (M \wedge T)).$$

If one thinks of "provided that" as indicating a sufficient condition, one has

$$((U \wedge L) \vee (M \wedge T)) \supset D.$$

Or if one thinks of the condition as both necessary and sufficient, one has

$$D \equiv ((U \wedge L) \vee (M \wedge T)).$$

The phrase "provided that" is used in all three ways, and of course the truth table depends on the interpretation chosen.

## Chapter 3, Section 5

4. Beggars are not always rich.

## Chapter 3, Section 6

1. FDNF $p$
   FCNF $p$
   Simplest $p$.

2. FDNF $(p \wedge q) \vee (\sim p \wedge q) \vee (\sim p \wedge \sim q)$
   FCNF $\sim p \vee q$
   Simplest $p \supset q$.

3. FDNF $(p \wedge q \wedge r) \vee (p \wedge \sim q \wedge \sim r) \vee (\sim p \wedge q \wedge \sim r)$
   $\vee (\sim p \wedge \sim q \wedge r)$
   FCNF $(p \vee q \vee r) \wedge (p \vee \sim q \vee \sim r) \wedge (\sim p \vee q \vee \sim r)$
   $\wedge (\sim p \vee \sim q \vee r)$
   Simplest $p \equiv q \equiv r$.

4. FDNF $(p \wedge \sim q \wedge r) \vee (p \wedge \sim q \wedge \sim r) \vee (\sim p \wedge q \wedge r)$
$\vee (\sim p \wedge q \wedge \sim r) \vee (\sim p \wedge \sim q \wedge r) \vee (\sim p \wedge \sim q \wedge \sim r)$

   FCNF $(\sim p \vee \sim q \vee r) \wedge (\sim p \vee \sim q \vee \sim r)$

   Simplest $p \mid q$.

5. FDNF $(p \wedge q \wedge r) \vee (p \wedge q \wedge \sim r) \vee (\sim p \wedge q \wedge r)$
$\vee (\sim p \wedge \sim q \wedge \sim r)$

   FCNF $(p \vee q \vee \sim r) \wedge (p \vee \sim q \vee r) \wedge (\sim p \vee q \vee r)$
$\wedge (\sim p \vee q \vee \sim r)$

   Simplest $(p \vee r) \equiv q$.

6. FDNF $(p \wedge \sim q \wedge r) \vee (p \wedge \sim q \wedge \sim r) \vee (\sim p \wedge q \wedge \sim r)$
$\vee (\sim p \wedge \sim q \wedge r) \vee (\sim p \wedge \sim q \wedge \sim r)$

   FCNF $(p \vee \sim q \vee \sim r) \wedge (\sim p \vee \sim q \vee r) \wedge (\sim p \vee \sim q \vee \sim r)$

   Simplest $q \mid (p \vee r)$.

7. FDNF $(p \wedge q \wedge r) \vee (p \wedge q \wedge \sim r) \vee (p \wedge \sim q \wedge r)$
$\vee (p \wedge \sim q \wedge \sim r) \vee (\sim p \wedge q \wedge r) \vee (\sim p \wedge q \wedge \sim r)$
$\vee (\sim p \wedge \sim q \wedge r) \vee (\sim p \wedge \sim q \wedge \sim r)$

   FCNF none

   Simplest $p \supset p$.

8. FDNF $(p \wedge q \wedge r) \vee (p \wedge q \wedge \sim r) \vee (p \wedge \sim q \wedge r)$
$\vee (\sim p \wedge q \wedge r) \vee (\sim p \wedge q \wedge \sim r) \vee (\sim p \wedge \sim q \wedge r)$

   FCNF $(p \vee q \vee r) \wedge (\sim p \vee q \vee r)$

   Simplest $q \vee r$.

9. FDNF $(p \wedge q \wedge r) \vee (p \wedge q \wedge \sim r) \vee (p \wedge \sim q \wedge r)$
$\vee (\sim p \wedge q \wedge r) \vee (\sim p \wedge q \wedge \sim r) \vee (\sim p \wedge \sim q \wedge r)$
$\vee (\sim p \wedge \sim q \wedge \sim r)$

   FCNF $\sim p \vee q \vee r$

   Simplest $p \supset (q \vee r)$.

10. FDNF $(p \wedge q \wedge r) \vee (p \wedge \sim q \wedge r) \vee (\sim p \wedge q \wedge r)$
$\vee (\sim p \vee \sim q \wedge r)$

   FCNF $(p \vee q \vee r) \wedge (p \vee \sim q \vee r) \wedge (\sim p \vee q \vee r)$
$\wedge (\sim p \vee \sim q \vee r)$

   Simplest $r$.

11. FDNF $(p \wedge q \wedge r) \vee (p \wedge \sim q \wedge \sim r) \vee (\sim p \wedge q \wedge r)$
$\vee (\sim p \wedge q \wedge \sim r) \vee (\sim p \wedge \sim q \wedge r)$

   FCNF $(p \vee q \vee r) \wedge (\sim p \vee q \vee \sim r) \wedge (\sim p \vee \sim q \vee r)$

   Simplest $(p \supset (q \equiv r)) \wedge (p \vee q \vee r)$.

12. FDNF $(p \wedge q \wedge r) \vee (p \wedge q \wedge \sim r) \vee (\sim p \wedge q \wedge r)$
$\vee (\sim p \wedge q \wedge \sim r) \vee (\sim p \wedge \sim q \wedge r)$
$\vee (\sim p \wedge \sim q \wedge \sim r)$

   FCNF $(\sim p \vee q \vee r) \wedge (\sim p \vee q \vee \sim r)$

   Simplest $p \supset q$.

13. FDNF    $(p \wedge q \wedge \sim r) \vee (\sim p \wedge \sim q \wedge \sim r)$
    FCNF    $(p \vee q \vee \sim r) \wedge (p \vee \sim q \vee r) \wedge (p \vee \sim q \vee \sim r)$
            $\wedge (\sim p \vee q \vee r) \wedge (\sim p \vee q \vee \sim r)$
            $\wedge (\sim p \vee \sim q \vee \sim r)$

    Simplest    $r \downarrow (p \not\equiv q)$ or $(p \equiv q) \not\supset r$.

14. FDNF    $(p \wedge q \wedge r \wedge s) \vee (p \wedge q \wedge \sim r \wedge s)$
            $\vee (p \wedge \sim q \wedge \sim r \wedge s)$
    FCNF    $(p \vee q \vee r \vee s) \wedge (p \vee q \vee r \vee \sim s)$
            $\wedge (p \vee q \vee \sim r \vee s) \wedge (p \vee q \vee \sim r \vee \sim s)$
            $\wedge (p \vee \sim q \vee r \vee s) \wedge (p \vee \sim q \vee r \vee \sim s)$
            $\wedge (p \vee \sim q \vee \sim r \vee s) \wedge (p \vee \sim q \vee \sim r \vee \sim s)$
            $\wedge (\sim p \vee q \vee r \vee s) \wedge (\sim p \vee q \vee \sim r \vee s)$
            $\wedge (\sim p \vee q \vee \sim r \vee \sim s) \wedge (\sim p \vee \sim q \vee r \vee s)$
            $\wedge (\sim p \vee \sim q \vee \sim r \vee s)$

    Simplest    $p \wedge (r \supset q) \wedge s$.

15. FDNF    $(p \wedge q \wedge r \wedge \sim s) \vee (p \wedge q \wedge \sim r \wedge s)$
            $\vee (p \wedge q \wedge \sim r \wedge \sim s) \vee (p \wedge \sim q \wedge r \wedge \sim s)$
            $\vee (\sim p \wedge q \wedge r \wedge s) \vee (\sim p \wedge q \wedge r \wedge \sim s)$
            $\vee (\sim p \wedge \sim q \wedge r \wedge \sim s)$
    FCNF    $(p \vee q \vee r \vee s) \wedge (p \vee q \vee r \vee \sim s)$
            $\wedge (p \vee q \vee \sim r \vee \sim s) \wedge (p \vee \sim q \vee r \vee s)$
            $\wedge (p \vee \sim q \vee r \vee \sim s) \wedge (\sim p \vee q \vee r \vee s)$
            $\wedge (\sim p \vee q \vee r \vee \sim s) \wedge (\sim p \vee q \vee \sim r \vee \sim s)$
            $\wedge (\sim p \vee \sim q \vee \sim r \vee \sim s)$

    Simplest    $(r \supset s) \mid (q \supset (p \equiv r))$ or $(r \supset s) \supset (q \wedge (p \not\equiv r))$.

## Chapter 3, Section 7

1. (a) A wff.
   (b) Not a wff.
   (c) A wff.
   (d) Not a wff.
   (e) Not a wff.
2. $C C p C p p p$.
3. $C K p N q E N N p q$.
4. $E E p q r$.
5. $C K p q K N q A p r$.
6. $E C N A K p q r p q$.
7. $E C p N q C r A N q p$.

8. $E E E p q r E p E q r$.

9. $C C A p r A p q A q r$.

10. $C p E A q r A p K N q r$.

11. $E A N p r C C p K q N q r$.

12. $E p E q E r E p N q$
   1 2 2 3 3 4 4 5 5 6

| p | q | r | formula |
|---|---|---|---------|
| — | — | T | F |
| — | — | F | T |

13. $A p C K N p N q K p E q r$
   1 2 2 3 4 5 4 5 3 4 4 5 5

| p | q | r | formula |
|---|---|---|---------|
| F | F | — | F |
| otherwise | | | T |

14. $C N N p A K N p N A N C E p q r A r N K r p q$
   1 2 3 4 2 3 4 5 4 5 6 7 8 9 9 8 6 7 7 8 9 9 3

| p | q | r | formula |
|---|---|---|---------|
| T | F | — | F |
| otherwise | | | T |

15. $A C A N K A E E p q r s p q r s$
   1 2 3 4 5 6 7 8 9 9 8 7 6 4 3 2

| p | q | r | s | formula |
|---|---|---|---|---------|
| — | T | F | F | F |
| F | F | F | F | F |
| otherwise | | | | T |

16. $K A N K A K E E p p p p q q q q$
   1 2 3 4 5 6 7 8 9 9 8 7 6 5 3 2

| p | q | formula |
|---|---|---------|
| — | T | T |
| — | F | F |

17. $A r C A N s q C A K E p q r p s$
   1 2 2 3 4 5 4 3 4 5 6 7 7 6 5 4

| p | q | r | s | formula |
|---|---|---|---|---------|
| T | — | F | F | F |
| otherwise | | | | T |

## Chapter 3, Section 8

2. (a) $p \lor q$ eq $\sim p \supset q$, $p \land q$ eq $\sim(p \supset \sim q)$
   (b) $\sim p$ eq $p \downarrow p$, $p \lor q$ eq $(p \downarrow q) \downarrow (p \downarrow q)$,
   $p \land q$ eq $(p \downarrow p) \downarrow (q \downarrow q)$.

3. We observe that $p \equiv \sim q$ eq $\sim(p \equiv q)$, and that $\equiv$ satisfies associativity and commutativity. Thus in any expression in $\equiv$ and $\sim$, we may move all negations to the outside of the expression, and associate terms from the left. If the expression contains at least three variable occurrences, the central grouping is essentially $(p \equiv p) \equiv q$, which is logically equivalent to $q$. Thus any expression in $\equiv$ and $\sim$, and in two variables $p$ and $q$ is logically equivalent to one of the following:

$$p, q, \sim p, \sim q, p \equiv p, p \equiv q, \sim(p \equiv p), \sim(p \equiv q).$$

But $p \wedge q$ is not equivalent to any of these.

## Chapter 3, Section 9

1. Show  $p \supset q, q \supset r, p \vdash r$

$p \supset q$

$p$

$q$

$q \supset r$

$r$

Then Deduction Theorem, once.

2. Show  $p \supset (q \supset r), q, p \vdash r$

$p \supset (q \supset r)$

$p$

$q \supset r$

$q$

$r$

Then Deduction Theorem, once.

3. $(\sim p \supset \sim \sim p) \supset ((\sim p \supset \sim p) \supset p)$

$\sim p \supset \sim p$

$\vdash (\sim p \supset \sim \sim p) \supset p$    (by Exercise 2).

4. $(\sim p \supset \sim \sim p) \supset p$

$\sim \sim p \supset (\sim p \supset \sim \sim p)$

$\vdash \sim \sim p \supset p.$

5. $(\sim \sim \sim p \supset \sim p) \supset ((\sim \sim \sim p \supset p) \supset \sim \sim p)$

$\sim \sim \sim p \supset \sim p$

$\vdash (\sim \sim \sim p \supset p) \supset \sim \sim p.$

6. $(\sim \sim \sim p \supset p) \supset \sim \sim p$

$p \supset (\sim \sim \sim p \supset p)$

$\vdash p \supset \sim \sim p.$

7. $\sim p \supset (\sim q \supset \sim p)$
   $\sim p$
   $\sim q \supset \sim p$
   $(\sim p \supset \sim p) \supset ((\sim q \supset p) \supset q)$
   $(\sim q \supset p) \supset q.$

8. Show    $\sim p \vdash p \supset q$
   $\sim p$
   $(\sim q \supset p) \supset q$    (by Exercise 7).
   $p \supset (\sim q \supset p)$
   $p \supset q$

Then Deduction Theorem, once.

9. Show    $\sim (p \supset \sim q) \vdash p$
   $\sim (p \supset \sim q)$
   $(\sim p \supset (p \supset \sim q)) \supset p$
   $\sim p \supset (p \supset \sim q)$
   $p$

Then Deduction Theorem, once, and definition of $\wedge$.

10. Show    $\sim (p \supset \sim q) \vdash q$
   $\sim (p \supset \sim q)$
   $(\sim q \supset (p \supset \sim q)) \supset q$
   $(\sim q \supset (p \supset \sim q))$
   $q$

Then Deduction Theorem, once, and definition of $\wedge$.

11. Show    $\sim q \supset \sim p, p \vdash q.$
   $p \supset \sim \sim p$
   $p$
   $\sim \sim p$
   $(\sim q \supset \sim p) \supset q$
   $(\sim q \supset \sim p)$
   $q$

Then Deduction Theorem, twice.

12. $p \supset q$
   $\sim \sim p \supset p$
   $\sim \sim p \supset q.$

13. $p \supset q$
   $\sim \sim p \supset q$
   $q \supset \sim \sim q$
   $\sim \sim p \supset \sim \sim q.$

14. Show    $p \supset q, \sim q \vdash \sim p$
   $\sim q$
   $(\sim \sim p \supset q) \supset \sim p$
   $p \supset q$

$\sim \sim p \supset q$

$\sim p$

Then Deduction Theorem, twice.

15. $(p \supset q) \supset q$

$((p \supset q) \supset q) \supset (\sim q \supset \sim(p \supset q))$

$\sim q \supset \sim(p \supset q)$.

16. Show  $p \vdash (\sim q \supset \sim(p \supset q))$

$(p \supset q) \supset (p \supset q)$

$p$

$(p \supset q) \supset q$

$\sim q \supset \sim(p \supset q)$

Then Deduction Theorem, once.

17. Show  $p \vdash q \supset \sim(p \supset \sim q)$

$p \supset (\sim \sim q \supset \sim(p \supset \sim q))$

$p$

$\sim \sim q \supset \sim(p \supset \sim q)$

$q \supset \sim \sim q$

$q \supset \sim(p \supset \sim q)$

Then Deduction Theorem, once, and definition of $\wedge$.

18. $\vdash \sim q \supset ((\sim p \supset q) \supset p)$     (by Exercise 7).

$\sim p \supset q$

$\sim q \supset p$

$(\sim q \supset \sim p) \supset ((\sim q \supset p) \supset q)$

$(\sim q \supset \sim p) \supset q$.

19. Show  $p \supset q, \sim p \supset q \vdash q$

$p \supset q$

$(p \supset q) \supset (\sim q \supset \sim p)$

$\sim q \supset \sim p$

$\sim p \supset q$

$(\sim q \supset \sim p) \supset q$

$q$

Then Deduction Theorem, twice.

## Chapter 4, Section 1

1. (a) $A$:  0 1 1 0 0 1 0 0 0 0

    $B$:  0 1 1 0 0 1 0 1 1 0

    $C$:  0 0 1 0 0 1 1 1 0 0

    $D$:  0 0 1 1 1 0 0 0 0 0

$E:$  0 1 0 0 0 0 1 1 0 1
$F:$  0 0 0 0 0 0 0 0 0 0
$G:$  1 1 1 1 1 1 1 1 1 1
(b) $R:$  0 1 0 0 0 0 0 0 0 0
$S:$  1 1 1 0 0 1 0 1 0 0
$T:$  0 1 1 1 1 1 0 1 1 0

## Chapter 4, Section 2

1. A:  1 1 1 0 0 0 0 0 0 0
   B:  0 0 1 1 1 1 0 0 0 0
   C:  1 1 0 0 0 0 1 0 0 0
   D:  1 0 0 0 1 0 0 1 0 0
   E:  1 0 0 0 1 1 0 0 1 0
   F:  0 0 1 1 0 1 0 0 0 1
   G:  0 0 0 0 0 0 1 0 1 0
   H:  1 1 0 0 1 0 0 0 0 0
   I:  0 0 0 1 0 1 0 0 1 0
   J:  1 0 0 1 0 0 0 1 0 0
   K:  0 0 1 0 0 0 1 0 0 0
   L:  1 1 0 0 1 0 0 0 0 0
   M:  0 0 0 0 0 1 0 0 1 1
   N:  0 0 1 1 0 0 0 0 1 1
   O:  0 0 0 1 0 0 0 1 1 0
   $\alpha:$  D and J
   $\beta:$  B, F, I, M, N
   $\gamma:$  no documents satisfy this request.

## Chapter 4, Section 3

1. (a) 155, (b) 28, (c) 21, (d) 5319, (e) 74, (f) 2.
2. (a) 10001, (b) 10111010100, (c) 10011101110101, (d) 11000000111001, (e) 11000, (f) 1010110100.
3. Set the product initially to 000, and let $A$ and $B$ represent the two numbers to be multiplied.

If the last (unit) digit of $B$ is 1, add $A$ to the product, using the addition method of the text.

Shift $A$ one to the left.

If the second digit of $B$ is 1, add the shifted $A$ to the product.

Shift $A$ one to the left.

If the first digit of $B$ is 1, add the shifted $A$ to the product.

## Chapter 5, Section 1

1. The device can add, subtract, multiply, divide, take square root, and test for zero or negative.

   (1) Compute $b^2 - 4ac$.

   (2) If $b^2 - 4ac$ is negative, go to step 7; otherwise continue.

   (3) Compute $\sqrt{(b^2 - 4ac)}$ and call it $d$.

   (4) Compute $(b + d)/2$: this is Answer 1.

   (5) Compute $(b - d)/2$: this is Answer 2.

   (6) Go to step 13.

   (7) Multiply $b^2 - 4ac$ by $-1$.

   (8) Compute $\sqrt{-(b^2 - 4ac)}$ and call it $e$.

   (9) Compute $b/2$.

  (10) Compute $e/2$.

  (11) Form (not compute!) $b/2 + ie/2$: this is Answer 1.

  (12) Form $b/2 - ie/2$: this is Answer 2.

  (13) Print Answer 1 and Answer 2, and halt.

2. Let the determinant entries be $a_{ij}$ where $i$ and $j$ take the values 1, 2, and 3. The device can add, subtract, and multiply.

The subscript computation is as follows: if $j = 3$, change $j$ to 1; otherwise increase $j$ by 1.

The computation of the determinants:

   (1) Set SUM equal to 0.

   (2) Set $j_1 = 1$, $j_2 = 2$, $j_3 = 3$.

   (3) Compute $p = a_{1j_1} a_{2j_2} a_{3j_3}$.

   (4) Add $p$ to SUM.

   (5) If $j_1 = 3$, go to step 8; otherwise continue.

   (6) Compute new values of $j_1$, $j_2$, and $j_3$.

   (7) Go to step 3.

   (8) Set $j_2 = 2$, $j_3 = 1$.

   (9) Compute $p = a_{1j_1} a_{2j_2} a_{3j_3}$.

  (10) Subtract $p$ from SUM.

  (11) If $j_1 = 2$, print SUM and halt; otherwise continue.

  (12) Compute new values of $j_1$, $j_2$, and $j_3$.

  (13) Go to step 9.

Note that in Step 8 we do not need to set $j_1 = 3$, since it already has that value. This example illustrates the independent specification of a subtask or *subroutine* which is frequently used in the problem solution; in this case the subscript computation.

3. The device can mark the paths in two ways to indicate the travel, say green for the paths which have been traversed once and red for the paths which have been covered twice. In addition, it has the ability to remember the last path segment covered, so that it can backtrack out of loops.

(a) If we are at point B, halt indicating success; otherwise continue.

(b) If we are at point A, go to step (l); otherwise continue.

(c) If there are three green lines at the point, go to step (j); otherwise continue.

(d) If there are unmarked lines at the point, go to step (h); otherwise continue.

(e) If there are green lines at the point, go to step (f); otherwise halt indicating failure.

(f) Choose a green line and follow it to the next point, marking it red.

(g) Go to step (a).

(h) Choose an unmarked line and follow it to the next point, marking it green.

(i) Go to step (a).

(j) Return along the last path segment covered, marking it red.

(k) Go to step (a).

(l) If there are two green lines at the point, go to step (j); otherwise go to step (d).

The method of choosing the lines in steps (f) and (h) must be specified if we are to have an algorithm.

4. Number the squares of the board in order, beginning in the upper left corner, then (1) put an $X$ in corner 1, and (2) if the opponent has marked corner 9, put an $X$ in corner 3; otherwise put an $X$ in corner 9.

These two instructions govern the first two moves, respectively. Subsequent moves are made according to the following rules:

(3) If there are two $X$'s in a line without an $O$, mark the third $X$ and halt with a win; otherwise continue.

(4) If there are two $O$'s in a line without an $X$, put an $X$ in the same line; otherwise continue.

(5) If corners 3 or 7 are unmarked, put an $X$ in one of these; otherwise continue.

(6) If there is an unmarked square, put an $X$ in it; otherwise halt with a draw.

(This is not necessarily the best algorithm for tic-tac-toe.)

**Chapter 5, Section 2**

1.

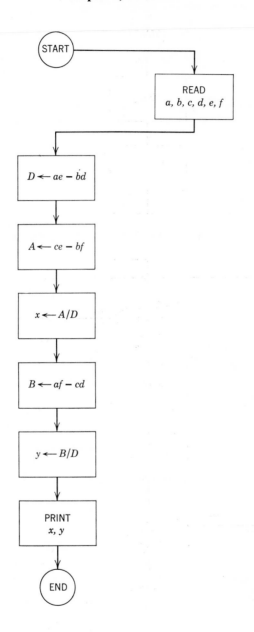

2.

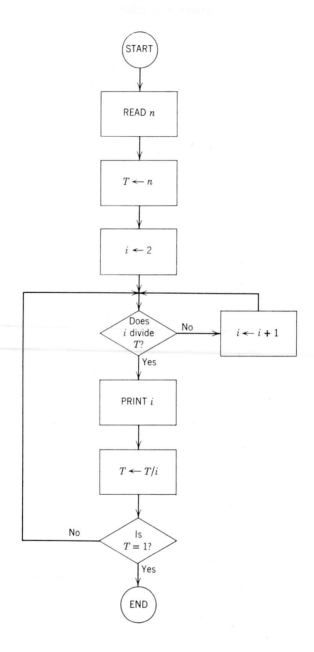

3.

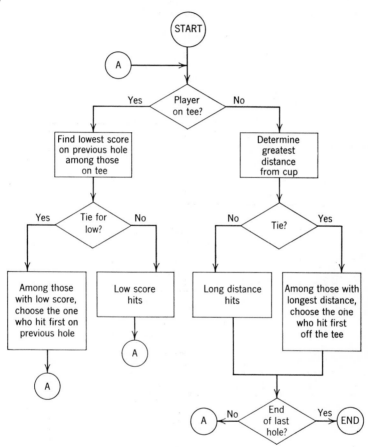

**4.**

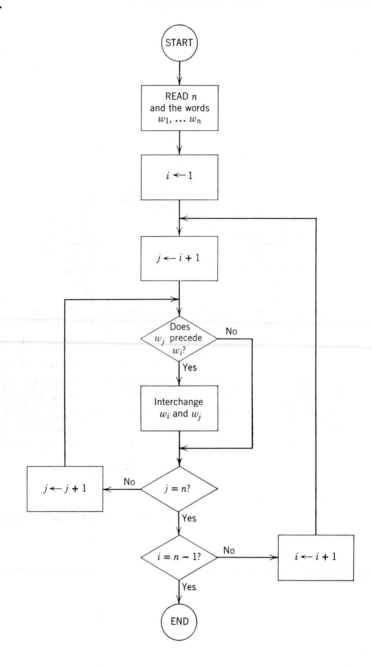

## Chapter 5, Section 3

1. $\Lambda \to .\Lambda$

2. $\Lambda \to \Lambda$

3. $a \to \Lambda$

4. $\alpha\zeta \to \alpha \qquad (\zeta \in \mathscr{A})$
   $\alpha \to .A$
   $\Lambda \to \alpha$

5. $\alpha\zeta \to \alpha \qquad (\zeta \in \mathscr{A})$
   $\alpha \to \Lambda$

6. $\zeta\alpha \to \alpha \qquad (\zeta \in \mathscr{A})$
   $\alpha \to \Lambda$
   $\beta\zeta \to \beta \qquad (\zeta \in \mathscr{A})$
   $\beta \to \Lambda$

7. $A \to .*A*$

8. Let $\mathscr{A} = \{a_1, a_2, \ldots, a_n\}$ and form a new isomorphic alphabet
   $\bar{\mathscr{A}} = \{\bar{a}_1, \bar{a}_2, \ldots, \bar{a}_n\}$.
   $\gamma\alpha \to \alpha\delta$
   $\gamma\zeta \to \bar{\zeta}\zeta\gamma \qquad (\zeta \in \mathscr{A})$
   $\delta\zeta \to \zeta\delta \qquad (\zeta \in \mathscr{A})$
   $\delta \to \beta$
   $\bar{\zeta}\eta \to \eta\bar{\zeta} \qquad (\zeta, \eta \in \mathscr{A})$
   $\bar{\zeta}\alpha \to \alpha\zeta \qquad (\zeta \in \mathscr{A})$
   $\bar{\zeta}\beta \to \beta\zeta \qquad (\zeta \in \mathscr{A})$
   $\epsilon \to .\Lambda$
   $\Lambda \to \epsilon\gamma$

9. If $n$ is represented by $n$ marks, use $+ \to \Lambda$

   If $n$ is represented by $n + 1$ marks (so that zero may be represented), use

   $$+ \to \Lambda$$
   $$1 \to .\Lambda$$

10. Assume that $n$ is represented by $n$ marks, and that $x$ and $y$ are strictly positive.

    $$\bar{1}\epsilon \to \epsilon 1$$
    $$\epsilon \to .\Lambda$$
    $$\delta 1 \to \delta$$
    $$\delta\bar{1} \to \bar{1}\delta$$
    $$\bar{1}\delta \to \epsilon 1$$
    $$\gamma 1 \to 1\bar{1}\gamma$$
    $$\gamma\bar{1} \to \bar{1}\gamma$$

$$\gamma \to \Lambda$$
$$\beta* \to *\gamma$$
$$\beta 1 \to 1\beta$$
$$\alpha 1 \to \alpha\beta$$
$$\alpha* \to \delta$$
$$\Lambda \to \alpha$$

11. Assume that $n$ is represented by $n$ marks.

$$\epsilon\zeta \to \zeta\epsilon \qquad (\zeta \in \{0, \ldots, 9\})$$
$$\epsilon\beta \to \beta\alpha$$
$$0\delta \to 1\epsilon$$
$$1\delta \to 2\epsilon$$
$$\vdots$$
$$8\delta \to 9\epsilon$$
$$9\delta \to \delta 0$$
$$\delta \to 1\epsilon$$
$$\beta\gamma \to \delta\beta$$
$$1\gamma \to \gamma 1$$
$$\alpha 1 \to 1\alpha$$
$$1\alpha \to \gamma$$
$$\beta\alpha \to .\Lambda$$
$$\Lambda \to \beta\alpha.$$

12. The algorithm doubles the string, but leaves one copy in reversed order.

## Chapter 5, Section 4

1.

|       | $\square$ | $\$$ | $a$ | $b$ | $x_1$ | $x_2$ | $\ldots$ | $x_n$ |
|-------|-----------|------|-----|-----|-------|-------|----------|-------|
| $q_0$ | $aRq_1$   | $bRq_3$ |     |     | $x_1Lq_6$ | $x_2Lq_6$ | | $x_nLq_6$ |
| $q_1$ | $aLq_2$   | $bRq_3$ | $aRq_1$ | | $x_1Lq_6$ | $x_2Lq_6$ | | $x_nLq_6$ |
| $q_2$ | $aRq_1$   | $bLq_3$ | $aLq_2$ | | $x_1Rq_5$ | $x_2Rq_5$ | | $x_nRq_5$ |
| $q_3$ | $aLq_4$   | $\$Lq_6$ | $aRq_3$ | $bRq_3$ | $x_1Rq_5$ | $x_2Rq_5$ | | $x_nRq_5$ |
| $q_4$ | $aRq_3$   | $\$Rq_5$ | $aLq_4$ | $bLq_4$ | $x_1Rq_5$ | $x_2Rq_5$ | | $x_nRq_5$ |
| $q_5$ | $\square Lq_6$ | $\$Lq_6$ | $aRq_5$ | $bRq_5$ | $x_1Lq_6$ | $x_2Lq_6$ | | $x_nLq_6$ |
| $q_6$ | $\square Lq_6$ | $\$Lq_8$ | $\square Lq_6$ | $\$Lq_7$ | $x_1Lq_6$ | $x_2Lq_6$ | | $x_nLq_6$ |
| $q_7$ | $\square Rq_8$ | $\$Lq_8$ | $\square Lq_7$ | | $x_1Lq_6$ | $x_2Lq_6$ | | $x_nLq_6$ |
| $q_8$ | $\square Rq_8$ | | | | | | | |

2. (a) Assume that $n$ is represented by $n$ ones

|       | $0$ | $1$ | $A$ | $B$ |
|-------|-----|-----|-----|-----|
| $q_0$ | $ALq_0$ | $1Rq_1$ | | |
| $q_1$ | $BLq_0$ | $1Rq_0$ | | |

(b)

| | □ | 1 | 2 | 3 | 4 | ⋯ | 9 | 0 | A | B |
|---|---|---|---|---|---|---|---|---|---|---|
| $q_0$ | $ALq_1$ | $1Rq_0$ | $2Rq_0$ | $3Rq_0$ | $4Rq_0$ | ⋯ | $9Rq_0$ | $0Rq_0$ | | |
| $q_1$ | | $1Rq_1$ | $2Rq_0$ | $3Rq_1$ | $4Rq_0$ | ⋯ | $9Rq_1$ | $0Rq_0$ | | $BLq_0$ |

3. (a) Assume that $n$ is represented by $n$ ones.

| | 0 | 1 |
|---|---|---|
| $q_0$ | | $0Rq_1$ |
| $q_1$ | $1Rq_0$ | $0Rq_2$ |
| $q_2$ | $1Rq_1$ | $0Rq_0$ |

(b)

| | □ | 0 | 1 | 2 | 3 | ⋯ | 8 | 9 |
|---|---|---|---|---|---|---|---|---|
| $q_0$ | $0Rq_3$ | $□Rq_0$ | $□Rq_1$ | $□Rq_2$ | $□Rq_0$ | | $□Rq_2$ | $□Rq_0$ |
| $q_1$ | $1Rq_3$ | $□Rq_1$ | $□Rq_2$ | $□Rq_0$ | $□Rq_1$ | ⋯ | $□Rq_0$ | $□Rq_1$ |
| $q_2$ | $2Rq_3$ | $□Rq_2$ | $□Rq_0$ | $□Rq_1$ | $□Rq_2$ | | $□Rq_1$ | $□Rq_2$ |

4.

| | a | b | c | d | □ | w | x | y | z |
|---|---|---|---|---|---|---|---|---|---|
| $q_0$ | $wRq_1$ | $xRq_3$ | $yRq_5$ | $zRq_7$ | $□Rq_{30}$ | | | | |
| $q_1$ | $aRq_1$ | $bRq_1$ | $cRq_1$ | $dRq_1$ | $□Rq_2$ | | | | |
| $q_2$ | $wLq_9$ | $bLq_{31}$ | $cLq_{31}$ | $dLq_{31}$ | $□Lq_{11}$ | $wRq_2$ | $xRq_2$ | $yRq_2$ | $zRq_2$ |
| $q_3$ | $aRq_3$ | $bRq_3$ | $cRq_3$ | $dRq_3$ | $□Rq_4$ | | | | |
| $q_4$ | $aLq_{11}$ | $xLq_9$ | $cLq_{31}$ | $dLq_{31}$ | $□Lq_{11}$ | $wRq_4$ | $xRq_4$ | $yRq_4$ | $zRq_4$ |
| $q_5$ | $aRq_5$ | $bRq_5$ | $cRq_5$ | $dRq_5$ | $□Rq_6$ | | | | |
| $q_6$ | $aLq_{11}$ | $bLq_{11}$ | $yLq_9$ | $dLq_{31}$ | $□Lq_{11}$ | $wRq_6$ | $xRq_6$ | $yRq_6$ | $zRq_6$ |
| $q_7$ | $aRq_7$ | $bRq_7$ | $cRq_7$ | $dRq_7$ | $□Rq_8$ | | | | |
| $q_8$ | $aLq_{11}$ | $bLq_{11}$ | $cLq_{11}$ | $zLq_9$ | $□Lq_{11}$ | $wRq_8$ | $xRq_8$ | $yRq_8$ | $zRq_8$ |
| $q_9$ | | | | | $□Lq_{10}$ | $wLq_9$ | $xLq_9$ | $yLq_9$ | $zLq_9$ |
| $q_{10}$ | $aLq_{10}$ | $bLq_{10}$ | $cLq_{10}$ | $dLq_{10}$ | | $aRq_0$ | $bRq_0$ | $cRq_0$ | $dRq_0$ |
| $q_{11}$ | | | | | $□Lq_{12}$ | $aLq_{11}$ | $bLq_{11}$ | $cLq_{11}$ | $dLq_{11}$ |
| $q_{12}$ | $aLq_{12}$ | $bLq_{12}$ | $cLq_{12}$ | $dLq_{12}$ | $□Rq_{13}$ | $aLq_{12}$ | $bLq_{12}$ | $cLq_{12}$ | $dLq_{12}$ |
| $q_{13}$ | $□Rq_{14}$ | $□Rq_{17}$ | $□Rq_{20}$ | $□Rq_{23}$ | | | | | |
| $q_{14}$ | $aRq_{14}$ | $bRq_{14}$ | $cRq_{14}$ | $dRq_{14}$ | $□Rq_{15}$ | | | | |
| $q_{15}$ | $aRq_{15}$ | $bRq_{15}$ | $cRq_{15}$ | $dRq_{15}$ | $□Rq_{16}$ | | | | |
| $q_{16}$ | $aRq_{16}$ | $bRq_{16}$ | $cRq_{16}$ | $dRq_{16}$ | $aLq_{26}$ | | | | |
| $q_{17}$ | $aRq_{17}$ | $bRq_{17}$ | $cRq_{17}$ | $dRq_{17}$ | $□Rq_{18}$ | | | | |
| $q_{18}$ | $aRq_{18}$ | $bRq_{18}$ | $cRq_{18}$ | $dRq_{18}$ | $□Rq_{19}$ | | | | |
| $q_{19}$ | $aRq_{19}$ | $bRq_{19}$ | $cRq_{19}$ | $dRq_{19}$ | $bLq_{26}$ | | | | |
| $q_{20}$ | $aRq_{20}$ | $bRq_{20}$ | $cRq_{20}$ | $dRq_{20}$ | $□Rq_{21}$ | | | | |

4.

| | $a$ | $b$ | $c$ | $d$ | $\square$ | $w$ | $x$ | $y$ | $z$ |
|---|---|---|---|---|---|---|---|---|---|
| $q_{21}$ | $aRq_{21}$ | $bRq_{21}$ | $cRq_{21}$ | $dRq_{21}$ | $\square Rq_{22}$ | | | | |
| $q_{22}$ | $aRq_{22}$ | $bRq_{22}$ | $cRq_{22}$ | $dRq_{22}$ | $cLq_{26}$ | | | | |
| $q_{23}$ | $aRq_{23}$ | $bRq_{23}$ | $cRq_{23}$ | $dRq_{23}$ | $\square Rq_{24}$ | | | | |
| $q_{24}$ | $aRq_{24}$ | $bRq_{24}$ | $cRq_{24}$ | $dRq_{24}$ | $\square Rq_{25}$ | | | | |
| $q_{25}$ | $aRq_{25}$ | $bRq_{25}$ | $cRq_{25}$ | $dRq_{25}$ | $dLq_{26}$ | | | | |
| $q_{26}$ | $aLq_{26}$ | $bLq_{26}$ | $cLq_{26}$ | $dLq_{26}$ | $\square Lq_{27}$ | | | | |
| $q_{27}$ | $aLq_{27}$ | $bLq_{27}$ | $cLq_{27}$ | $dLq_{27}$ | $\square Lq_{28}$ | | | | |
| $q_{28}$ | $aLq_{29}$ | $bLq_{29}$ | $cLq_{29}$ | $dLq_{29}$ | | | | | |
| $q_{29}$ | $aLq_{29}$ | $bLq_{29}$ | $cLq_{29}$ | $dLq_{29}$ | $Rq_{13}$ | | | | |
| $q_{30}$ | | | | | | $aRq_{30}$ | $bRq_{30}$ | $cRq_{30}$ | $dRq_{30}$ |
| $q_{31}$ | | | | | $\square Lq_{32}$ | $aLq_{31}$ | $bLq_{31}$ | $cLq_{31}$ | $dLq_{31}$ |
| $q_{32}$ | $aLq_{32}$ | $bLq_{32}$ | $cLq_{32}$ | $dLq_{32}$ | | $aLq_{33}$ | $bLq_{33}$ | $cLq_{33}$ | $dLq_{33}$ |

## Chapter 5, Section 5

1.

| | 0 | 1 |
|---|---|---|
| $q_0$ | $1Rq_1$ | |

2.

| | 0 | 1 |
|---|---|---|
| $q_0$ | $1Rq_1$ | $1Lq_1$ |
| $q_1$ | $1Lq_0$ | |

3.

| | 0 | 1 |
|---|---|---|
| $q_0$ | $1Rq_1$ | $1Lq_2$ |
| $q_1$ | $1Rq_2$ | $1Rq_3$ |
| $q_2$ | $1Lq_0$ | $0Lq_1$ |

(This is just one of the possible solutions to the exercise.)

## Chapter 5, Section 6

1.
```
CLA   A
SUB   B
TRP   AA
TRZ   AB
CLA   A
LDQ   B
STQ   A
STO   B
```

```
AA   CLA   A
     DIV   B
     TRZ   AB
     LDQ   B
     STQ   A
     STO   B
     TRA   AA
AB   (print instruction—the result is in B).
```

2. Assume that the amount is given in cents, and that the locations HALF, QUART, DIME, NICK hold the constants 50, 25, 10, and 5 respectively.

```
          CLA   AMT
          DIV   HALF
          STQ   N(1)
          DIV   QUART
          STQ   N(2)
          DIV   DIME
          STQ   N(3)
          DIV   NICK
          STQ   N(4)
          STO   N(5)
```

The necessary number of coins are in $N(1) \ldots N(5)$.

3. Assume that INC and FAM hold the income and the number of family members, respectively, and that the following constants are present:

```
SIX      600
FIVE     5000
ONE      0.20
TWO      0.25
THREE    0.30
TEN      0.10
SML      1000
MED      2250
```

```
CLA   FAM
MPY   SIX
STQ   TEMP
CLA   INC
MPY   TEN
CLA   MQ
ADD   TEMP
STO   TEMP
CLA   INC
```

|     | SUB | TEMP |
|-----|-----|------|
|     | STO | INC |
|     | SUB | FIVE |
|     | TRP | AA |
|     | TRZ | AB |
|     | ADD | FIVE |
|     | MPY | ONE |
|     | STQ | TAX |
|     | TRA | END |
| AB  | CLA | SML |
|     | STO | TAX |
|     | TRA | END |
| AA  | SUB | FIVE |
|     | TRP | AC |
|     | TRZ | AD |
|     | ADD | FIVE |
|     | MPY | TWO |
|     | CLA | MQ |
|     | ADD | SML |
|     | STO | TAX |
|     | TRA | END |
| AD  | CLA | MED |
|     | STO | TAX |
|     | TRA | END |
| AC  | MPY | THREE |
|     | CLA | MQ |
|     | ADD | MED |
|     | STO | TAX |
| END | (print instruction—the result is in TAX). | |

## Chapter 6, Section 1

1. Term:              b, g, h, k
   Atomic:            e, j, m
   Nonatomic wff:     c, d, f, i, q
   None of these:     a, l, n, o, p
2. (a) $(x) (G(x) \supset H(x))$
   (b) $(x) ((W(x) \land F(x)) \supset (\exists y)(B(y) \land C(y, x)))$
   (c) $(x) (E(x) \supset ((C(x) \land D(x)) \lor (D(x) \land W(x))))$
   (d) For this one needs a pairing relationship, $M(x, y)$ ($x$ meets $y$) and an identity relation, $I(x, y)$ ($x$ is identical to $y$).

$(x)((B(x) \land P(x)) \supset (\exists y)(G(y) \land P(y) \land M(x, y)))$
$\land\ (x)(y)(z)(w)(B(x) \land P(x) \land B(z) \land P(z) \land G(y) \land P(y)$
$\land\ G(w) \land P(w) \land M(x, y) \land M(z, w) \land \sim I(x, z) \supset$
$\sim I(y, w))$.

(e) $(x)((B(x) \land P(x)) \supset (\exists y)(G(y) \land P(y) \land M(x, y))$
$\land\ (y)(z)(G(y) \land P(y) \land M(x, y) \land G(z) \land P(z)$
$\land\ M(x, z) \supset I(y, z))) \land (x)(y)(z)(w)$
$(B(x) \land P(x) \land B(z) \land P(z) \land G(y) \land P(y) \land G(w) \land P(w)$
$\land\ M(x, y) \land M(z, w) \land \sim I(x, z) \supset \sim I(y, w))$.

3. (a) For every number $x$ there is a number $y$ such that if $x$ is less than $y$ then for any number $z$, $z$ lies between $x$ and $y$.

(b) For any individuals $x$ and $y$ either $x$ and $y$ are unrelated, or $x$ is a father and the father of $x$ and $y$ is a grandfather.

(c) Either there does not exist an $x$ less than $y$, or every $y$ is less than $x$.

(d) For all $x$ and $y$, $x$ equals $y$ if and only if $y$ equals $x$.

(e) For all $x$ and $y$, $x$ is less than $y$, and there exists a $z$ such that either $x$ is less than $z$ or some $x$ is less than $y$.

## Chapter 6, Section 2

1. (a) The first two $x$'s are bound, the third one free. Both $y$'s are bound.

(b) $x$:   free, bound, free
$y$:   free, free, bound, bound
$z$:   bound, bound.

(c) $x$:   bound, bound, free
$y$:   bound, bound, bound, bound, free.

(d) $x$:   bound
$y$:   bound
$z$:   free, free.

(e) $x$:   free, bound, bound
$y$:   free, free, bound.

2. (a) $(x)A(x) \supset (\exists y)B(f(x, z), y)$.

(b) $A(x, g(y, z)) \land (\exists x)B(g(y, z)) \supset (y)(z)C(x, y, z)$ (The substitution for $x$ is not legal.)

(c) $(x)(\exists y)(A(y, x) \land (y)C(y)) \supset B(y, f(x, y))$.

(d) Not legal.

(e) $A(f(y)) \supset (B(f(y)) \supset (\exists x)(C(f(y)) \supset (y)D(x)))$.

## Chapter 6, Section 3

1. For $\mathscr{F}_1$ and $\mathscr{F}_3$, let the constant $b$ be any element of the domain and let the constant $a$ be the same as $b$ or any element preceding $b$. Then the formulas are valid. $\mathscr{F}_2$ is valid for any assignment of elements to $a$ and $b$.

2. (a) True, true.
   (b) True, true.
   (c) False, true.
   (d) True, true.
   (e) True, true.
   (f) False, true.

### Chapter 6, Section 4

1. Satisfiable for one individual, satisfiable for two.
2. Satisfiable, satisfiable.
3. Satisfiable, satisfiable.
4. Satisfiable, satisfiable.
5. Satisfiable, satisfiable.

### Chapter 6, Section 5

1. $(y)[\mathscr{A}(y) \lor \mathscr{B}]$.
2. $(\exists y)[\mathscr{A}(y) \lor \mathscr{B}]$.
3. $(y)[\mathscr{A}(y) \land \mathscr{B}]$.
4. $(\exists x)(z)[(\mathscr{A}(x) \supset \mathscr{B}) \land (\mathscr{B} \supset \mathscr{A}(z))]$.
5. $(\exists y)[\mathscr{A}(y) \mid \mathscr{B}]$.
6. $(\exists y)(\exists z)[A(y) \supset B(x, z)]$.
7. $(x_1)(y_1)(z_1)[(A(x, y) \land B(x_1)) \supset C(x, y_1, z_1)]$.
8. $(\exists z)(x_1)(\exists y_1)[(A(x_1, z) \land C(y_1)) \supset B(x, y)]$.
9. $A(z) \supset B(z)$.
10. $A(x) \supset (B(y) \supset (C(y) \supset D(x)))$.

### Chapter 6, Section 6

1. $(x)(y) \sim \mathscr{A} \equiv (y)(x) \sim \mathscr{A}$
   $\sim(x)(y) \sim \mathscr{A} \equiv \sim(y)(x) \sim \mathscr{A}$
   $\sim(x) \sim \sim(y) \sim \mathscr{A} \equiv \sim(y) \sim \sim(x) \sim \mathscr{A}$
   $(\exists x)(\exists y)\mathscr{A} \equiv (\exists y)(\exists x)\mathscr{A}$.
2. Show  $(x)(\mathscr{A} \supset \mathscr{B}), (x)\mathscr{A} \vdash (x)\mathscr{B}$
   $\qquad\qquad (x)(\mathscr{A} \supset \mathscr{B})$
   $\qquad\qquad \mathscr{A} \supset \mathscr{B}$
   $\qquad\qquad (x)\mathscr{A}$
   $\qquad\qquad \mathscr{A}$
   $\qquad\qquad \mathscr{B}$
   $\qquad\qquad (x)\mathscr{B}$
   Then Weak Deduction Theorem, twice.

3. Show    $(x)(\mathscr{A} \supset \mathscr{B}) \vdash (x)(\sim \mathscr{B} \supset \sim \mathscr{A})$
$(x)(\mathscr{A} \supset \mathscr{B})$
$\mathscr{A} \supset \mathscr{B}$
$(\mathscr{A} \supset \mathscr{B}) \supset (\sim \mathscr{B} \supset \sim \mathscr{A})$
$\sim \mathscr{B} \supset \sim \mathscr{A}$
$(x)(\sim \mathscr{B} \supset \sim \mathscr{A})$

Then Weak Deduction Theorem, once.

4. Show    $(x)(\mathscr{A} \supset \mathscr{B}) \vdash (\exists x)\mathscr{A} \supset (\exists x)\mathscr{B}$
$(x)(\mathscr{A} \supset \mathscr{B}) \supset (x)(\sim \mathscr{B} \supset \sim \mathscr{A})$
$(x)(\mathscr{A} \supset \mathscr{B})$
$(x)(\sim \mathscr{B} \supset \sim \mathscr{A})$
$(x)(\sim \mathscr{B} \supset \sim \mathscr{A}) \supset ((x) \sim \mathscr{B} \supset (x) \sim \mathscr{A})$
$(x) \sim \mathscr{B} \supset (x) \sim \mathscr{A}$
$\sim (\exists x)\mathscr{B} \supset \sim (\exists x)\mathscr{A}$     (from definition of $(\exists x)$)
$(\sim (\exists x)\mathscr{B} \supset \sim (\exists x)\mathscr{A}) \supset ((\exists x)\mathscr{A} \supset (\exists x)\mathscr{B})$
$(\exists x)\mathscr{A} \supset (\exists x)\mathscr{B}$

Then Weak Deduction Theorem, once.

5. Show first    $(x)(\mathscr{A} \wedge \mathscr{B}) \vdash (x)\mathscr{A} \wedge (x)\mathscr{B}$
$(x)(\mathscr{A} \wedge \mathscr{B})$
$\mathscr{A} \wedge \mathscr{B}$
$\mathscr{A}$
$\mathscr{B}$
$(x)\mathscr{A}$
$(x)\mathscr{B}$
$(x)\mathscr{A} \wedge (x)\mathscr{B}$

Hence by the Weak Deduction Theorem, $(x)(\mathscr{A} \wedge \mathscr{B}) \supset ((x)\mathscr{A} \wedge (x)\mathscr{B})$.
The same proof, read from bottom up, shows that $(x)\mathscr{A} \wedge (x)\mathscr{B}) \supset$
$(x)(\mathscr{A} \wedge \mathscr{B})$. Then use the definition of $\equiv$.

6. $(x)(\mathscr{A} \wedge \sim \mathscr{B}) \equiv (x)\mathscr{A} \wedge \sim (\exists x)\mathscr{B}$
Then use the definition of $\wedge$.

## Chapter 7, Section 1

1. Alphabet a, f, x, F, l, $*$, $\mathscr{V}$, $\mathscr{C}$, $\mathscr{F}$, $\mathscr{P}$
   Axioms $\mathscr{V}$x1, $\mathscr{C}$a1, $\mathscr{F}$f1$*$1, $\mathscr{P}$F1$*$1
   Productions   $\mathscr{V}\alpha \rightarrow \mathscr{V}\alpha 1$
   $\mathscr{C}\alpha \rightarrow \mathscr{C}\alpha 1$
   $\mathscr{F}\alpha *\beta \rightarrow \mathscr{F}\alpha 1*\beta$
   $\mathscr{F}\alpha *\beta \rightarrow \mathscr{F}\alpha *\beta 1$
   $\mathscr{P}\alpha *\beta \rightarrow \mathscr{P}\alpha 1*\beta$
   $\mathscr{P}\alpha *\beta \rightarrow \mathscr{P}\alpha *\beta 1$

2. (a) *a, ba, bab, bbb, dcbba.*
   (b) *bab, dcbaa, badcba.*

(c) *bc, dabb, dcdaba, badcdab.*

(d) *badcdab, dcdabbbdcda.*

(e) *bab, badcba, adcbbb.*

(f) *badcdab, adcdbbbdcda.*

(g) *a, ba, bab, dcbaa, badcba, bc, dabb, bbb, dcbba, badcbb, adcbbbdcb, dcadcbbbdca.*

(h) *bc, dabb, dcdaba, badcdab, adcdbbbb, a, ba, bab, bbbdcdbbb.*

(i) *a, ba, bc, dabb, bbb, dcbba, badcbb, bab, bbbdcb.*

(j) *bbbdcb, dcbbbdca, badcbbbdc, dabadcbbbdb.*

## Chapter 7, Section 2

1. Alphabet:  0,1,2,3,4,5,6,7,8,9,.,S,I,D,T

   Axioms:  S0,S1,S2,S3,S4,S5,S6,S7,S8,S9,S.,D0,D1,D2,D3,D4,D5,
   D6,D7,D8,D9

   Productions:  $D\alpha \rightarrow I\alpha$

   $D\alpha, I\beta \rightarrow I\alpha\beta$

   $I\alpha \rightarrow T\alpha$

   $I\alpha \rightarrow T\alpha.$

   $I\alpha, I\beta \rightarrow T\alpha.\beta$

2. $\langle$Art$\rangle$ ::= a | an | the

   $\langle$Adj$\rangle$ ::= colorless | green | old | square

   $\langle$N$\rangle$ ::= ball | ideas | John

   $\langle$V$\rangle$ ::= is | plays | sleep | sleeps

   $\langle$Adv$\rangle$ ::= furiously | quietly

   $\langle$AP$\rangle$ ::= $\langle$Adj$\rangle\langle$N$\rangle$ | $\langle$Adj$\rangle\langle$AP$\rangle$

   $\langle$NP$\rangle$ ::= $\langle$Art$\rangle\langle$AP$\rangle$ | $\langle$Adj$\rangle\langle$AP$\rangle$ | $\langle$N$\rangle$

   $\langle$Pred$\rangle$ ::= $\langle$Adv$\rangle$

   $\langle$VP$\rangle$ ::= $\langle$V$\rangle$ | $\langle$V$\rangle\langle$Pred$\rangle$ | $\langle$V$\rangle\langle$NP$\rangle$

   $\langle$S$\rangle$ ::= $\langle$NP$\rangle\langle$VP$\rangle$

# Index